Statistics and Experimental Design

Statistics and Experimental Design

AN INTRODUCTION FOR BIOLOGISTS AND BIOCHEMISTS

Geoffrey M. Clarke

M.A., Dip Stat (Oxon), C. Stat

A member of the Hodder Headline Group
LONDON
Distributed in the United States of America by
Oxford University Press Inc., New York

First published in Great Britain 1969
Second edition 1980
Third edition 1994 by Edward Arnold
a member of the Hodder Headline Group
338 Euston Road, London NW1 3BH

Distributed in the United States of America by
Oxford University Press Inc.,
198 Madison Avenue, New York, NY10016

British Library Cataloguing in Publication Data
A catalogue record for this book is available from the British Library.

Library of Congress Cataloging-in-Publication Data
A catalog record for this book is available from the Library of Congress.

ISBN 0 340 59324 5

9 10

Typeset in Times by Computape (Pickering) Ltd, North Yorkshire
Printed and bound in Malta

Preface

In spite of the ever-increasing presence of statistics in school and college courses, many students of the biological sciences still have the same problems as in the past with mathematical and numerical work. In particular there is difficulty in thinking in symbols, so that the statistics needed for biological work must be learnt through an 'applied' approach, seeing the use of methods and the reasons for them. In fact this is a very good way of learning statistics, because many gross errors in applying statistical methods arise through not knowing when they work, and through using methods that are not at all appropriate to the problem that is being studied. Arithmetical errors are, by comparison, probably less prevalent or serious these days because calculations will often be carried out using a computer.

This text aims to present the ideas behind, and the limitations of, standard statistical procedures as well as the details of the methods themselves. It has to be admitted that people do sometimes appear to find this logical thinking about methods just as hard as appreciating what the symbols in the standard formulae mean! But there is no substitute for thinking before using any scientific method, statistical or otherwise; and indeed if we are going to make numerical measurements as the main end-point of an experiment or the main records in a survey we have to decide carefully and precisely what these shall be as part of the detailed planning that should go into every good piece of scientific work.

In biology it is often possible to plan the conduct of field or laboratory experiments in considerable detail, and so emphasis has been placed in this book on the elementary aspects of experimental design; often if some basic statistical principle of design has been overlooked it will not be possible to carry out a fully satisfactory analysis, and the value of much work and effort by the experimenter can be lost. To study the basic principles of experimental design and analysis helps in understanding some of the commonly-used tests of significance, most of which arose originally from such situations.

The rapid developments of user-friendly computers allow students and workers in the biological sciences to study their data before carrying out the arithmetical routine of a significance test or other statistical calculation. This study may be in the form of graphs, diagrams or tables, and if carried out using a good program it should help a great deal in deciding whether particular methods are suitable for the data being examined. There is no substitute for taking a careful look at data before putting them through some standard computer program: we have only ourselves to blame if we obtain meaningless bits of arithmetic out of a program because that analysis was inappropriate for telling us anything about those data. When using the common, standard statistical distributions as models for sets of data, it is very important to check whether the conditions in which a particular distribution applies are satisfied, at least as a pretty good approximation, in our situation.

The original text with this title appeared in the Contemporary Biology series and went through two editions. The present book is a complete revision of that text. It acknowledges the place of computers in studying and analysing data in two ways: first, it includes some advice on using MINITAB in statistical work, as this program is widely available for teaching purposes; and second, it gives prominence to graphical displays of data. So-called Initial Data Analysis depends largely on visual and simple numerical methods. When these had to be done by hand they could be rather tedious and so were not often used; they are, however, very informative in deciding what further analyses should be attempted and in checking the conditions for these further analyses to be valid.

Another change from the previous text is the greater space given to worked examples, and the fuller explanation and discussion of results, made possible by the larger page size. In addition to these worked examples, there are answers and comments, collected at the end of the book, for the Exercises that follow the chapters. A student working alone would be well advised to attempt all (or most) of these Exercises before moving on to each new chapter. References to the theoretical basis of the methods, and to a few of the relevant computing problems, are provided from other books published by Edward Arnold: *A Basic Course in Statistics*, by G. M. Clarke and D. Cooke, and *Basic Statistical Computing*, by D. Cooke, A. H. Craven and G. M. Clarke.

Many of the principles of elementary statistical methods apply quite generally over a wide range of sciences, and it is hoped that the book may continue to be of use to scientists in disciplines other than biology, provided they do not mind most of the examples being from biology and biochemistry.

I am indebted to the Biometrika Trustees for permission to reprint parts of Tables 8, 12, 13 and 18 from *Biometrika Tables for Statisticians*, third edition; also to the Literary Executor of the late Sir Ronald Fisher, F.R.S., to Dr. Frank Yates, F.R.S., and to Oliver & Boyd Ltd., Edinburgh, for permission to reprint parts of Tables III, V and VI from their book *Statistical Tables for Biological, Agricultural and Medical Research*, sixth edition.

I would like to acknowledge the help and encouragement received from Professor Arthur Willis as General Editor of the Contemporary Biology series when the earlier texts were written, and his continuing interest in the subject. Comments from a wide variety of users have been useful and will continue to be very welcome.

North Chailey, Sussex, 1993.   G.M.C.

Contents

Symbols

r_i	a typical value of a discrete variate r
x_i	a typical value of a continuous variate x
f_i	the frequency with which x_i occurs in a sample
$\bar{x}$ or $\bar{r}$	the mean of a sample of observations
n	the number of observations in a sample
M	the median of a sample of observations
q	the lower quartile of a sample of observations
Q	the upper quartile of a sample of observations
s^2	the variance of a sample of observations
μ, σ^2	mean and variance of a probability distribution
$N(\mu, \sigma^2)$	the normal distribution whose parameters are μ, σ^2
n, p	the parameters in a binomial distribution
q	$1 - p$
$\hat{p}$	estimate of p calculated from a sample
λ	the parameter (mean) of a Poisson distribution
$\hat{\lambda}$	estimate of λ calculated from a sample
N.H.	Null Hypothesis upon which the calculations of a significance test are based
A.H.	Alternative Hypothesis, accepted when N.H. is rejected
d.f.	degrees of freedom
$t_{(f)}$	Student's t distribution with f degrees of freedom
$\chi^2_{(f)}$	the χ^2-distribution with f degrees of freedom
$F_{(m, n)}$	the F (variance-ratio) distribution with m and n d.f.
ρ	correlation coefficient in a population
r	estimate of ρ calculated from a sample
r_s	Spearman's rank correlation coefficient
U	the Mann-Whitney test statistic
b	slope of regression line
$\hat{b}$	estimate of b calculated from a sample
z	the unit (standard) normal deviate
S.S.	sum of squares of deviations about the mean
M.S.	mean square (S.S./d.f.)
$\hat{\sigma}^2$	estimate of variance calculated in analysis of variance
e_{ij}	error term in a linear (analysis of variance) model

List of illustrations

1 Natural variation and types of data

1.1 Introduction

Experiments in physical science often aim to estimate a numerical *constant* such as the acceleration due to gravity from an experiment using a simple pendulum. If someone obtains an answer very different from the known true value, he or she has made an error at some stage in the experiment, perhaps not using the equipment carefully enough. Again, if the relation between voltage and current is studied in an electrical circuit, the resistance in the circuit must be kept constant; if it is allowed to vary, and is not controlled, the voltage–current relationship will not be discovered. In controlled conditions, it is commonly supposed in physical and chemical experiments that there is one correct answer, not subject to noticeable variation.

Immediately units of plant or animal material, or human subjects, are used in an experiment, however well-controlled the environmental conditions, another very important factor arises: *natural variation*. Two plants, grown originally from the same batch of seed in identical pots side by side on a greenhouse bench, and given the same amount of fertiliser and water over the same period of time, will not make exactly the same growth. Twin animals of the same sex, living in the same environment and receiving the same diet, will not put on exactly the same weight in a given time. In fact, natural variation exists in many other areas of scientific experimentation besides biology, but it was discovered first by biologists because it is usually large enough to notice even in very carefully controlled work. Later, industrial experimenters found that two specimens, cut from the same sheet of metal and treated alike, will not have exactly the same breaking strength; two pieces of cloth cut from the same production roll will not have exactly the same number of faults per metre; two car tyres made from the same supply of rubber will not last exactly the same lifetime. The purity of two samples of a pharmaceutical product will vary somewhat even when they are taken from the same batch. Therefore the methods we describe in this text can be applied much more widely than biology and biochemistry. The common factor is that experimental units, or plants, or animals *vary among themselves naturally even when they are treated alike*. This variation came to be called 'experimental error', which is not a good name for it since it is not usually error in the sense of 'mistake'. We may also use the names 'random variation' or 'residual variation' to describe this natural variation.

Statistics is that branch of mathematics which studies natural variation. Suppose that we have a number of individuals which are of the same sort—that is, they come from the same *population*. It is very important to specify this population precisely and unambiguously; there would be little point in studying the variation in a group of individuals all basically different from one another, because we should not then be looking at natural random variation. These individuals would all differ systematically, for reasons which we could usually specify: two plants may be of different cultivars (varieties), two animals of different sex or parentage, two people carrying out a sequence of computer operations may have different amounts of experience with the equipment. A **population**, in the scientific statistical sense, consists of items, or units, or individuals all having the same characteristics and existing in the same controlled conditions or environment. Within this population, the natural variation may often show a recognisable pattern; when it does so, there may be standard methods of extracting information from the numerical data we can collect on the population. We shall examine some of these methods later.

Usually populations are very large: we could go on growing plants from the same batch of seed, then another generation from the seed of those, then a third generation and so on; or we could go on mass-producing industrial items by the same process without limit. Often we shall want to say something about the population of plants or the population of that industrial product. But these statements about a population will have to be based on experiments using only a few units, a *sample* from the whole population. The sample must give a good representation of the population if the statements ('*inferences*') we make are to be of any use. We return to this point when designing and planning experiments; the sample must usually be a *random sample* from the whole population (see Section 4.3).

1.2 Types of data

1.2.1 Example
The times, measured to the nearest second, taken by 30 people to carry out a sequence of computer operations are given below. Before this test, they had all received the same training in keyboard skills and the use of that particular computer. During the test, conditions were kept as standard as possible.

47	46	48	42	43	46
61	68	54	48	51	44
53	48	41	65	45	52
43	72	63	45	38	43
46	57	49	44	58	47

What do we mean by quoting a time to the nearest second? In fact, if we had very accurate equipment, we could measure times to decimals of a second, so the record '47 seconds' does not necessarily mean *exactly* 47. By the usual scientific rules for rounding numbers to the nearest unit, 47 will actually mean 'from 46·5 up to, but not including, 47·5 seconds'. Time is a *continuous* measurement, one which can take any value whatever, to any number of decimal places. Our record of it is only limited by the available equipment.

1.2.2 Example
The numbers of male rats, in litters containing five rats altogether, from a closed population are shown on page 3:

Numbers of male rats
2, 3, 3, 2, 0, 3, 2, 3, 2, 1, 2, 3, 0, 5, 1, 4, 2, 4, 1, 3, 2, 2, 3, 4, 4.

There is no question of decimals or fractions here: the number can only be 0, 1, 2, 3, 4 or 5. This is not a continuous measurement—it is usually called *discrete*—and we can write down a list of all the possible values which the recorded number may take.

1.2.3 Example
Both the sets of data given above may be called *quantitative*: they are numerical measurements of some quantity which we hope will give useful information.

Another type of data is *qualitative*; an observation is made which is not a measurement. Suppose that a genetic characteristic which can be observed depends on two genes *A*, *B*, and these can appear in either dominant (*A*, *B*) or recessive (*a*, *b*) form. Then four types of member will appear in the population: *AB*, *Ab*, *aB*, *ab*. A qualitative observation simply records which of these types each member is. A list made from studying a sample of ten may look like this:

Ab, ab, AB, AB, aB, AB, Ab, aB, AB, AB.

1.3 Summarising data

The three examples in Section 1.2 are typical of data collected in different types of scientific experiment or observation, but as they stand it is difficult to see any pattern in them. The first step in summarising data is to draw a suitable type of graph or diagram. Computers with graphical display make this easy to do for large sets of data, although not all commonly-used statistical packages do all the most useful forms of display. Another summary is to make a *frequency table*, counting up the total frequency or number of occurrences of each value 0, 1, 2, 3, 4, 5 in Example 1.2.2, the frequency of each of the four genetic types in Example 1.2.3, or the number of items in each *interval* 35–38, 39–42, and so on up to 67–70, 71–74 seconds in Example 1.2.1.

The third, very important, form of summary is to calculate an 'average' value of all the data in the sample. The arithmetic mean is a very common summary measure, but there are others; different types of summary suit different types of data. These numerical ways of summarising data are very useful when two different populations or samples of measurements are being compared in an experiment. In an experiment on plant growth, for example, one set of the plants will receive a standard fertiliser treatment and the other set will have additional nitrogen; apart from this difference the growing conditions and environment will be controlled as closely as possible to remain constant. The effect of additional nitrogen will be assessed by comparing the average heights of the two sets of plants. We return to this topic in Chapter 3.

From now on in this text, we shall assume that the study of a set of data always begins with a suitable graphical display. At present the package MINITAB is widely available for teaching purposes in Statistics laboratories and we shall refer to it occasionally. But any package locally available (provided it has a user-friendly instruction manual!) will do many of the same tasks. It is a serious part of a statistical analysis to decide what methods are suitable for studying the data available at the end of a piece of experimental work. Simply putting these data through some well-known computer routine will often miss a lot of the information that the experiment has given, either because one or two of the observations have some particularly interesting individual characteristics or because that computer routine is not suitable for those data.

We shall also assume that when one of the standard statistical distributions is used as a model for our data the first step is to check off all the standard conditions that need to be satisfied; as we introduce each distribution we shall emphasise the conditions required. These conditions are just as much biological as statistical; neither the biologist nor the statistician working together on a project can leave it wholly to the other to make this check.

2 Studying and summarising data graphically

2.1 Graphical methods for studying data

The aim of displaying data in graphs, charts and diagrams is to discover any patterns present in the data. Also we can extend graphical methods so as to compare two or more sets of data. We will concentrate on methods most often required in biology; there are others which are very useful for demographic, economic and social data.

For *discrete data*, all possible values of the measurement or record are listed together with the *frequency* (total number of occurrences) of each in the sample. Example 1.2.2 gives the *frequency table*:

No. of male rats	0	1	2	3	4	5	Total
Frequency	2	3	8	7	4	1	25

This information is shown in a *bar diagram*, Fig. 2.1. The possible values of the record 0, 1, 2, 3, 4, 5 are marked on the horizontal axis and the frequencies are represented by (thick) vertical lines at each value.

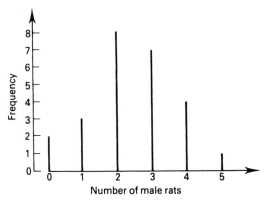

Fig. 2.1   Numbers of male rats in litters of five rats.

A space must be left between 0 and 1, and 1 and 2, etc.; if this is not done, the impression of a *discrete* measurement is lost. (A bar diagram must not be confused with a histogram—see Section 2.2.)

An alternative is a *dot-plot*, which is often provided by a computer graph-plotter program. Each observation is represented by a dot in the correct vertical column. Figure 2.2 shows a dot-plot for the same data, and provided care is taken to place the dots the same distance apart vertically, the heights of the columns of dots give the same information as the bars in Fig. 2.1.

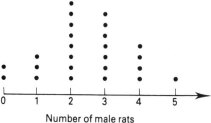

Number of male rats

Fig. 2.2 Dot-plot of rat data.

Both of these methods of display help to show the pattern of frequencies for the different possible values of the measurement; for example, they show that the middle values 2 and 3 are much the most common, while the extremes, 0 and 5, are quite rare.

2.2 Continuous data

Sets of continuous data with not too many observations, such as in Example 1.2.1, may be shown by a *stem-and-leaf* diagram. In the simplest form of stem and leaf, for two figure numbers like these, the tens form the 'stems' and units the 'leaves', recorded in order as they are read from the list of data. The vertical line separates the stems (3, 4, 5, 6, 7) from the leaves. Reading from 1.2.1, the first entry is 47 and appears in the second row of Fig. 2.3 as 4|7. The next data item is 46, and this also goes on the second row to make 4|76. Going on through the set of data in this way, Fig. 2.3 is completed.

```
3 | 8
4 | 7682368481 53536947
5 | 413278
6 | 1853
7 | 2
```

Fig. 2.3 Stem-and-leaf diagram of computing time data.

If the leaves are recorded neatly in columns, we can compare the frequencies in each stem.
    Another refinement of a stem-and-leaf diagram is to arrange the leaves on each stem in ascending order as in Fig. 2.4: this is often useful in making numerical summaries (Section 3.2).

```
3 | 8
4 | 123334455666778889
5 | 123478
6 | 1358
7 | 2
```

Fig. 2.4 Ordered stem-and-leaf diagram of computing times.

A very common way of representing continuous data is in a *histogram*. This requires more skill in programming to give a genuine computer output histogram, and some programs which are called 'histogram' actually give something more like a dot-plot or a stem-and-leaf plot (see Section 2.6).

First, we must decide what width of interval the data shall be grouped into. The decision will affect the resulting histogram somewhat, and may even give apparently different shape. There should be enough intervals to help show any pattern in the frequencies, but not so many that frequencies become very small. If we use 35–38, 39–42, . . . up to 71–74 seconds, we have the frequency table:

Interval	35–38	39–42	43–46	47–50	51–54	55–58	59–62	63–66	67–70	71–74	Total
Frequency	1	2	10	6	4	2	1	2	1	1	30

There are too many intervals to give a useful picture (there are too many small frequencies), although with large samples of data 10 intervals will often be suitable. One possible rule is to use about $\sqrt{n}$ intervals when there is a total of n observations. Since the values in Example 1.2.1 range from 38 up to 72, a distance of 34 units, and there are $n = 30$ observations altogether, the number of intervals should be about five or six and the width will then be six or seven units. Choosing six intervals, each six units wide, gives the frequency table:

Interval	38–43	44–49	50–55	56–61	62–67	68–73	Total
Frequency	6	13	4	3	2	2	30

Remembering that these are data recorded to the nearest whole number, the intervals are from 37.5 up to but not including 43·5; 43·5–49·5; 49·5–55·5; 55·5–61·5; 61·5–67·5; 67·5–73·5. Any value recorded as '·5' goes in the upper of the two intervals by the usual rounding rules.

A histogram (Fig. 2.5) represents the frequency in each interval by the *area* of a rectangle standing on that interval. While intervals are all the same length, as here, the heights are therefore proportional to the frequencies; but in general intervals need not be the same length and then the areas must be made to show the frequencies.

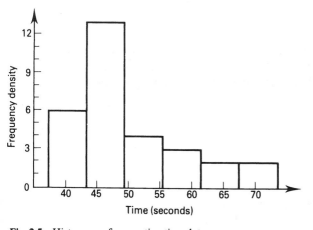

Fig. 2.5   Histogram of computing time data.

A dot-plot may also be used for continuous data, but is less helpful when the data take a wide range of values (see Fig. 2.6).

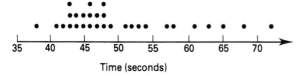

Fig. 2.6   Dot-plot for computing time data.

2.2.1 Example

The following data are the survival times of 125 greenfly in a study of the effects of an insecticide:

Survival time (seconds)	0–20	20–40	40–60	60–100	100–140	140–240
Number of insects	10	51	32	19	9	4

Present these data in a histogram.

In this example, the time-intervals are not equal, because most of the insects had been affected in the first 60 seconds and so it would not give very useful information to go on recording the few remaining so accurately. For the histogram, the vertical scale is not frequency but *frequency density*, and as usual we should remember that by the rounding rule the end-points of the intervals are 19·5, 39·5, 59·5, 99·5, 139·5. Areas standing on each interval must represent total frequencies in the intervals. Taking 20 seconds as the unit for interval width, the height of the rectangle for '20–40' will be 51, and for '40–60' it will be 32. Because '60–100' is an interval two units wide, the height must be half the frequency, $\frac{1}{2} \times 19 = 9\cdot5$; similarly for '100–140' the height is 4·5, and in '140–240' we must take the height $\frac{1}{5} \times 4 = 0\cdot8$ because the interval is five units wide. Strictly speaking, the first interval is only 19·5 seconds wide, not 20; i.e. it is not quite one unit but $\frac{19\cdot5}{20} = 0\cdot975$ units. The height of the rectangle representing frequency should then be $\frac{10}{0\cdot975} = 10\cdot256$, instead of 10. Obviously the times cannot go below 0! But this refinement is not always included.

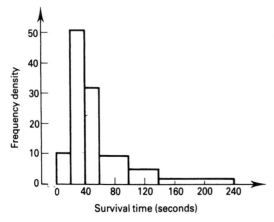

Fig. 2.7   Histogram of survival times of greenfly.

2.3 Choice of intervals for stem-and-leaf diagrams

The same problem of showing patterns can arise in stem-and-leaf diagrams as in histograms. The diagrams in Section 2.2 would be better if 'stems' of less than 10 were chosen. It is usually easy to adapt a program to give stems of 5 instead of 10, and where data are recorded to one decimal place instead of the nearest whole number, this can be very useful (stems being the whole numbers and leaves the decimals, arranged in two groups, ·0 to ·4 and ·5 to ·9, for each whole number). Figure 2.8 gives considerably more information than Fig. 2.4, on the same data.

```
3   |
·   | 8
4   | 1233344
·   | 55666778889
5   | 1234
·   | 78
6   | 13
·   | 58
7   | 2
·   |
```

Fig. 2.8 Alternative stem-and-leaf plot for computing time data.

2.4 Cumulative frequency tables and diagrams

When data are arranged in a frequency table, in increasing order of size of the measurement recorded, the *cumulative frequencies* may be shown on an extra row of the table. The cumulative frequency for any value X is the total frequency of all the observations having values less than or equal to X. Thus for Example 1.2.2 we have:

No. of male rats	0	1	2	3	4	5
Frequency	2	3	8	7	4	1
Cumulative frequency	2	5	13	20	24	25

For discrete data, this is usually all that is needed.

For continuous data, grouped into suitable intervals in a frequency table, a *cumulative frequency diagram* is useful (sometimes called an ogive because of its general shape). For Example 1.2.1 grouped into six intervals, as in Section 2.2, the complete table is:

Interval	38–43	44–49	50–55	56–61	62–67	68–73
Frequency	6	13	4	3	2	2
Cumulative frequency	6	19	23	26	28	30

This means that there are six observations *up to 43·5*, 19 altogether up to 49·5, and so on, remembering what the actual end-points of the intervals are. A *cumulative frequency diagram* (or *curve*) – see Fig. 2.9 – plots these cumulative frequencies against the end-points of the intervals. These points are joined freehand by as smooth a curve as possible. When this curve is drawn very carefully it is a good way of finding some of the numerical summary measures for a set of data, particularly the median and quartiles (Section 3.2). Its starting point is the bottom of the lowest interval, here 37·5.

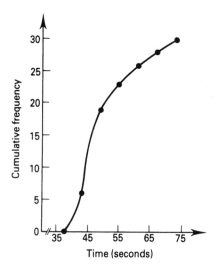

Fig. 2.9   Cumulative frequency curve of computing time data.

The same principle will be applied to a set of frequencies shown in unequal intervals, such as in Example 2.2.1. That gives the table

Survival time (secs)	0–20	20–40	40–60	60–100	100–140	140–240
Frequency	10	51	32	19	9	4
Cumulative frequency	10	61	93	112	121	125

and using the interval end-points already explained, the cumulative frequency curve is shown in Fig. 2.10.

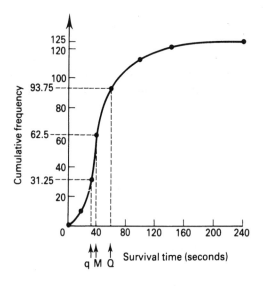

Fig. 2.10   Cumulative frequency curve of greenfly survival times.

2.5 Comparing two sets of data

Two sets of discrete data, where the measurements are both in similar size regions (e.g. one set ranges from 0 to 10 and the others from 5 to 20) can be compared using dot-plots one below the other on the same sheet of paper using the complete scale (e.g. 0 to 20). Alternatively, bar diagrams may be drawn one below the other in a similar way. Stem-and-leaf diagrams can be used in the same way for continuous data, except that if possible (provided neither data set is too large) they should be side-by-side on the paper for easier comparison. The aim of comparison is to see whether the two patterns are similar in shape, whether one is more spread out than the other, whether one is like the other, except for being on a different part of the scale of measurement. Comparisons like these are what we aim to make numerically when we carry out statistical significance tests. The characteristics to look for are:

location—where the 'centre' is;
dispersion—how scattered the data are;
skewness—symmetrical or not.

These are the most important characteristics of any statistical distribution. If data are roughly symmetrical they are easy to handle mathematically; if two data sets are of similar shape they are easy to compare using standard statistical procedures. Careful use of diagrams should help to avoid some of the worst misuses of statistical methods.

2.6 Displays in MINITAB

The instruction *Stem-and-Leaf* produces exactly the same displays as we have seen in Section 2.2; it gives an ordered diagram like Fig. 2.4. The distance between stems can be set using the *increment* instruction, and can be 1, 2 or 5, with leading or trailing zeros, such as 10, 20 or 50; 0.1, 0.2 or 0.5.

Also a *Dotplot* (out of the *Graph* menu) produces just that (e.g. Fig 2.2)

Recent releases of MINITAB have made it possible to obtain histograms, either with equal intervals (like Fig. 2.6) or unequal intervals (like Fig. 2.7). First select *Histogram* from the *Graph* menu, then in *Graph Variables* give the number of the column containing that data which are to be plotted on the horizontal axis. From the *Data Display* dialog box choose *Bar*. MINITAB chooses intervals of equal length and marks their midpoints on the *x*-axis.

In order to specify our own choice of intervals, we must select *Options* in the Histogram option window, select *Cutpoint* for *Type of Intervals,* and finally in *Definition of Intervals* click *Midpoint/cutpoint positions.* This allows the boundaries (i.e. the ends) of the intervals to be typed in.

When intervals are **not** of equal width, the frequency plot has to be replaced by a frequency **density** plot (as explained on page 8). Boundaries of intervals can be specified as above, but for *Type of Histogram* we have to select *Density*.

2.7 Exercises on Chapter 2

2.7.1 The number, r, of males in litters of rats which contain five rats altogether, was observed with the following results over 100 different litters. Summarise these results and illustrate them: r = 2, 5, 3, 1, 3, 4, 2, 2, 0, 3, 2, 4, 1, 3, 3, 2, 2, 2, 1, 2, 3, 4, 3, 0, 4, 5, 3, 3, 3, 2, 3, 2, 2, 1, 1, 4, 1, 3, 2, 2, 4, 3, 4, 3, 2, 3, 2, 2, 3, 4, 2, 3, 0, 1, 1, 3, 5, 2, 2, 4, 3, 2, 5, 4, 3, 2, 4, 3, 3, 4, 2, 1, 3, 2, 1, 2, 4, 3, 1, 3, 3, 4, 2, 3, 1, 2, 3, 4, 3, 2, 2, 4, 3, 3, 4, 1, 2, 3, 2, 3.
[This is the full set of data from which Example 1.2.2 came.]

2.7.2 The weights (in g) of 120 animals, of similar age and genetic history, at the end of a spell of feeding on the same diet in controlled conditions, were

Weight x:	50–80	80–90	90–100	100–110	110–120
Frequency f:	3	6	13	25	24

Weight x:	120–130	130–150	150–180	180–240
Frequency f:	21	18	7	3

Prepare a suitable diagram to illustrate these observations.

2.7.3 The numbers of the taxonomic group mollusca counted in 50 sampling units (of fixed size) at each of two sites in the Lewes Brooks area were:

Site A: 2, 1, 3, 0, 1, 0, 6, 2, 8, 0, 9, 10, 6, 0, 4, 1, 2, 3, 22, 3, 0, 1, 0, 6, 1, 3, 8, 0, 4, 4, 0, 1, 2, 5, 1, 4, 1, 0, 2, 3, 10, 5, 8, 0, 0, 6, 0, 4, 6, 0

Site B: 7, 22, 0, 4, 4, 2, 0, 4, 1, 4, 0, 0, 6, 1, 0, 2, 1, 4, 6, 3, 0, 3, 2, 6, 3, 5, 2, 0, 0, 1, 5, 1, 0, 5, 3, 0, 1, 0, 15, 4, 3, 0, 4, 4, 1, 2, 0, 1, 0, 0

Summarise and compare these two sets of data.

2.7.4 The weights in grams of 60 seedlings of a plant when taken from a greenhouse for planting out were:

21, 18, 33, 40, 20, 19, 38, 47, 26, 32, 15, 31, 33, 28, 42, 37, 40, 29, 24, 25, 15, 38, 31, 36, 34, 25, 28, 29, 22, 42, 29, 29, 40, 35, 32, 36, 34, 41, 27, 31, 45, 24, 27, 40, 19, 16, 45, 15, 42, 28, 18, 31, 19, 33, 25, 26, 20, 36, 27, 37

Summarise these in suitable tables and diagrams.

2.7.5 The heights (in mm) of 187 plants growing in a greenhouse from the same source of seed and sown at the same time are as follows. (f_x represents the frequency of measurement x).

x:	26	27	28	29	30	31	32	33	34	35	36	37	38	39
f_x:	1	0	2	1	0	1	3	0	1	0	4	1	2	6
x:	40	41	42	43	44	45	46	47	48	49	50	51	52	53
f_x:	10	12	8	6	15	17	20	13	9	12	7	8	8	6
x:	54	55	56	57	58	59	60	61	62	63	Total			
f_x:	4	4	0	1	1	0	1	1	1	1	187			

(a) Make a frequency table using intervals of width 5 mm, starting at 26 mm. Draw a histogram and a cumulative frequency curve.

(b) Repeat, using intervals of width 4 mm, and compare the two histograms. Which, if either, gives more information?

3 Numerical summaries of data

3.1 Frequency polygon and frequency curve

Exercise 2.7.5 gave details of the heights of 187 similar plants growing in the same environment, and in addition to the graphical display suggested there we may also write down a frequency table whose first row contains the mid-points of the intervals. Since records were to the nearest mm, the intervals are '26–30', '31–35' etc., and their actual end-points will thus be 30·5, 35·5, . . . so that mid-points are 28, 33, 38, . . .

Mid-point	28	33	38	43	48	53	58	63
Frequency in interval	4	5	23	58	61	30	3	3

Note that this is a *continuous* variate, despite the discrete look to the table; the frequency 61, for example, includes *all* the observations between 45·5 and 50·5, i.e. everything recorded 46, 47, 48, 49 or 50.

We may now plot a graph showing mid-points in the x direction and corresponding frequencies as y (Fig. 3.1, the crosses). A *frequency polygon* is obtained by joining these plotted points (the dotted line in Fig. 3.1). We expect a continuous variate to behave in a fairly smooth manner, and so the best possible smooth curve is drawn over this polygon (the *frequency curve* in Fig. 3.1, shown by the continuous line).

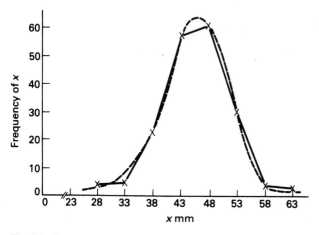

Fig. 3.1 Frequency polygon (straight lines) and frequency curve (dotted) for the data of Exercise 2.7.5 as summarised in the frequency table.

Sometimes, a certain amount of guesswork is needed in the drawing, and in smoothing out the sharp corners of the frequency polygon. In this example, the behaviour at the peak is not clear, and in fact a slightly different picture would emerge if the widths of the class-intervals were changed.

In the same way, when a diagram such as a histogram, dot-plot or stem-and-leaf diagram has been drawn to illustrate a set of continuous data, it is common to put a smooth curve over the diagram to show the general shape. A stem-and-leaf diagram shows the same information as a histogram but turned on its side, i.e. through 90°: the pattern of frequencies comes from the ends of the rows instead of the tops of the rectangles. The shape of the continuous *frequency curve* can then be summarised in the same way as we have already mentioned, using the three basic characteristics of *location*, *dispersion* and *skewness*. Numerical summaries of data aim to describe these three.

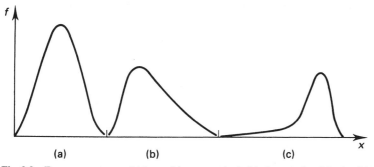

Fig. 3.2   Frequency curves which are (a) symmetrical, (b) skew to the right (positively skew), (c) skew to the left (negatively skew).

Figure 3.2 shows different types of curve which may be found with biological data: in (b) there are a small number of larger values, some way removed from the bulk of the data; in (c)—which is rare—the situation is reversed; in (a) the pattern is symmetrical.

Figure 3.3 indicates how we shall aim to compare two (or more) distributions. As regards location, two curves may be of the same shape but occupy different positions on the *x*-axis, because the measurements of *x* for the population represented by one curve are larger than those for the other; compare (α) and (β) in Fig. 3.3. For dispersion or scatter, two curves may again be of the same general shape, but one may contain *x* values extending over a much wider range than the other; compare (β) and (γ) in Fig. 3.3.

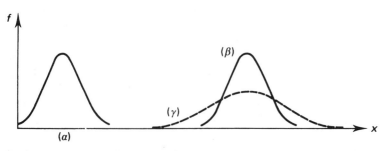

Fig. 3.3   (α) and (β): two symmetrical frequency distribution curves having the same variance but different means; (β) and (γ) two symmetrical curves having the same mean but different variances.

Before calculating any numerical measures of location and dispersion it is essential to see if the distributions have the same general shape. There is little that could usefully be said in comparing (a) and (c) in Fig. 3.2, for example; it would be best to summarise each one separately. All too often, routine calculations of numerical measures of location and dispersion are made without checking that the distributions involved are of (at least approximately) the same shape. When computers with graphics are available, this preliminary check can often be combined with the subsequent numerical work all in the same run, without wasting time.

3.2 Median and quartiles

The first step when dealing with small sets of data is to arrange them in increasing order of size: *ranking*, as it is called. This may be combined with producing a stem-and-leaf diagram (or a dot-plot). The computing time data in Example 1.2.1 did not give a very symmetrical pattern, but appeared 'skew to the right' (where the tail is). Other properties of these data are measured using *median* and *quartile* values.

Consider first a very simple set of data.

3.2.1 Example
The numbers of micro-organisms in unit volumes of water from a pond are:

$$10, 16, 12, 5, 22, 14, 19.$$

Arranged in order these are: 5, 10, 12, 14, 16, 19, 22.
The **median** (M) is the middle one in this ranked order:

$$5, 10, 12, 14, 16, 19, 22.$$
$$\uparrow$$
$$M$$

The observation 14 has three smaller and three larger, so is obviously the 'middle' in that sense. The median is a measure of *location*, or a measure of central tendency of the data. It is influenced only by the value(s) in the centre, not by the values at the ends. The median of 6, 7, 8, 14, 35, 45, 55 would also be 14, even though the data set is a very different shape from Example 3.2.1.

If there are eight observations, not seven, the extra one being 38, the data set of 3.2.1 becomes

$$5, 10, 12, 14, 16, 19, 22, 38.$$

Now there is a middle *pair*, not a middle one: 14 and 16. We shall place the median midway between these, and say $M = 15$. The extra value 38, although it is much larger than the next one (22), has made little difference to the median.

The median divides data into two parts: the *quartiles* divide data into four parts. One quarter of all the observations have values less than or equal to the lower quartile (q), one quarter between q and M, one quarter between M and the upper quartile (Q) and one quarter equal to or greater than Q. For Example 3.2.1, we find q and Q by finding the middle of the lower set (5, 10, 12) which is 10, and the middle of the upper set (16, 19, 22), which is 19. Thus $q = 10$ and $Q = 19$.

Again it is less straightforward with an even number of observations: Example 3.2.1 augmented by the observation 38 splits into two sets of four observations (5, 10, 12, 14) and (16, 19, 22, 38). We have placed M midway between 14 and 16. Similarly we place q midway between 10 and 12, so $q = 11$; and Q is midway between 19 and 22, so $Q = 20.5$.

The positions of q, M, Q help to show whether a set of data is symmetrical or not. If the distance $(M-q)$ is roughly equal to the distance $(Q-M)$ then the middle part of the data, at least, is fairly symmetrical. We return to this in Section 3.4 (box-and-whisker diagrams).

3.2.2 Example
The weight gains (g) of two sets of animals under different diets are:

(A)	56	67	42	48	55	61	52	39	47	58	50	40	59	62	44	57
(B)	78	34	37	72	58	68	27	55	65	40	75	33	66			

These data may be compared by dot-plots, using the same scale of measurement with one set of dots below the other. Neither set in this example contains any repeated values, so the picture is very clear and easy to see: set B is much more spread out than set A (Fig. 3.4).

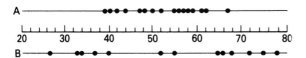

Fig. 3.4   Dot-plots to compare two sets of data.

An alternative comparison is to construct a stem-and-leaf plot for each set, and to place these side by side using the same scale (Fig. 3.5).

```
A 2 |                      B 2 | 7
  3 | 9                      3 | 347
  4 | 02478                  4 | 0
  5 | 0256789                5 | 25
  6 | 127                    6 | 568
  7 |                        7 | 258
```

Fig 3.5   Stem-and-leaf comparison of two data sets.

We would expect to find the same information from calculating the median and quartiles for each set of data. Set A consists of 16 observations, so the value of the median is midway between the eighth and ninth from the beginning in the rank order of size. The dot-plot can be used to find these two observations: their values are 52 and 55 so $M = 53.5$. In set B, there are 13 observations and so the seventh in the rank order is the median: its value is 55.

To find quartiles, set A is divided into four sets of four:

39,	40,	42,	44,	47,	48,	50,	52,	55,	56,	57,	58,	59,	61,	62,	67
			↑					↑				↑			
			q					M				Q			

The first quartile is midway between 44 and 47, and the third quartile is midway between 58 and 59; i.e. $q = 45.5$ and $Q = 58.5$.

Set B contains an odd number of observations, which means that M is exactly at one of them. We now need to split the lower half of the data into two equal parts to find q, and likewise the upper half for Q. The values are $q = 35.5$ and $Q = 70.0$.

27, 33, 34, 37, 40, 52, 55, 65, 66, 68, 72, 75, 78		
$\uparrow$ $\qquad\qquad\qquad\qquad$ M $\qquad\qquad\qquad$ $\uparrow$		
q $\qquad\qquad\qquad\qquad\qquad\qquad\qquad\qquad\qquad$ Q		

Using a cumulative frequency curve such as Fig. 2.10, the median and quartiles can be estimated. In Fig. 2.10, the total frequency is 125 and so the median must be the survival time corresponding to a frequency of 62.5; in the same way q is the time corresponding to a frequency 31.25, and Q to 93.75. These are estimated as approximately $q = 32$, $M = 41$ and $Q = 60$ seconds.

3.3 The arithmetic mean

The criticism of the median as a measure of location is that it uses only the middle values in the distribution. If we want a measure that uses them all, the *arithmetic mean* may be calculated. This is simply:

$$\frac{sum\ of\ all\ observations}{total\ number\ of\ observations}.$$

For a small set of data, such as Example 3.2.1, this follows immediately:

$$mean = (10 + 16 + 12 + 5 + 22 + 14 + 19)/7$$
$$= 98/7 = 14.0.$$

(Means rarely work out to whole numbers, even when the data are, and a useful rule is then to calculate a mean to one place of decimals—in general, to one more place of decimals than the original data.)

Let us see the effect of including the observation 38; now

$$mean = (10 + 16 + 12 + 5 + 22 + 14 + 19 + 38)/8$$
$$= 136/8 = 17.0.$$

As it happens, this is again a whole number; but what is more important is that it has been more seriously affected by including the extra observation than the median was. For the median, all that mattered was that one more observation had come in; for the mean the size of that observation is important. In a skew distribution, the median is a more stable central measure than the mean.

If there was something odd about that extra observation, and it was not really part of the same population at all, our estimate of the central location is much more upset if the mean is used than if the median had been. We cannot, of course, simply reject unusual-looking observations—so-called *outliers*—unless there is sound biological reason for doing so, and if possible a 'suspect' observation should be checked immediately while the experimental material is still available.

When there is a frequency table of *discrete* data already available, the mean can be found quickly from it. Example 1.2.2 was summarised as:

Number of males	0	1	2	3	4	5	Total
Frequency	2	3	8	7	4	1	25

The mean is $(0+0+1+1+1+2+\ldots+4+4+4+4+5)/25$ or

$$\frac{(2 \times 0) + (3 \times 1) + (8 \times 2) + (7 \times 3) + (4 \times 4) + (1 \times 5)}{25}$$

which is $(0+3+16+21+16+5)/25 = 61/25 = 2\cdot44$. (We have broken the one-decimal-place rule here because the answer happens to be exact to two places.)

After *continuous* data have been grouped into a grouped frequency table, and we no longer have the original observations, we can use a similar method once a 'typical' value has been taken to represent each interval. For Example 1.2.1, when intervals of six units were used, we can add *mid-points*, as in Section 3.1:

Interval	38–43	44–49	50–55	56–61	62–67	68–73	Total
Mid-point	40·5	46·5	52·5	58·5	64·5	70·5	
Frequency	6	13	4	3	2	2	30

Here 40·5 is midway between 37·5 and 43·5, which are the ends of that interval; and so on. The mean from the table is:

$$\frac{(6 \times 40\cdot5) + (13 \times 46\cdot5) + (4 \times 52\cdot5) + (3 \times 58\cdot5) + (2 \times 64\cdot5) + (2 \times 70\cdot5)}{30} = \frac{1503}{30} = 50\cdot1.$$

If we go back to the original data in Example 1.2.1, the total of the 30 observations is 1507, and the mean is therefore $1507/30 = 50\cdot23$. It is very unlikely that we shall find exacly the same mean value from the table and from the original data, because when grouping the data into intervals we lose some information—we no longer know exactly how the individual data were placed in each interval.

3.4 Box-and-whisker diagrams

A useful way to plot a large amount of information about a set of data is a box-and-whisker plot (so called by J. W. Tukey who suggested it). This shows the positions of median, quartiles and both extreme values in the data set. Example 3.2.1, with data ranging from 5 to 22, would be shown as in Fig. 3.6. When the extra observation 38 is included it takes the form in Fig. 3.7.

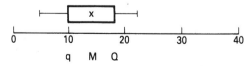

Fig. 3.6   Box-and-whisker diagram for Example 3.2.1.

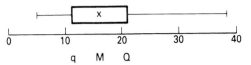

Fig. 3.7 Box-and-whisker diagram for augmented data of Example 3.2.1.

At once it can be seen that the extra observation makes the picture less symmetrical. There is a long 'whisker' at the top end when the value 38 is included, so a skew-to-the-right appearance has come in. The 'box', determined by the two quartiles, has moved to the right slightly, and so has the median, shown by the cross. But the general shape in this middle section is hardly altered. Box-and-whisker diagrams can be very helpful in comparing two or more sets of data before any further statistical analysis is attempted. They should of course be drawn to the same scale and placed underneath one another for easy comparison.

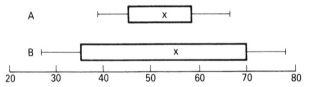

Fig. 3.8 Box-and-whisker plots for the data of Example 3.2.2.

The two sets of data from Example 3.2.2, on the weight gains of animals, contain similar numbers of animals and so it is reasonable to compare them by drawing two box-and-whisker plots on the same scale underneath one another, as in Fig. 3.8. This shows very simply what we have already noticed, that set B is much more spread out than set A; although the two medians take similar values, the distance between lower and upper quartiles (which is called the *inter-quartile range*) is much greater for B than for A.

3.5 Sigma-notation

The formula for the mean can be written down much more simply by using a mathematical notation and the greek capital Σ. If we had a simple set of data like Example 3.2.1, we could label up the values, in the order they were actually recorded, as $x_1, x_2, x_3, x_4, x_5, x_6, x_7$; thus $x_1 = 10$, $x_2 = 16$, up to $x_7 = 19$. A typical one of these is labelled x_i, and the sum of these is written $\sum_{i=1}^{7} x_i$, showing that i has the values from 1 to 7, all of which are to be included in the sum. Therefore $x_1 + x_2 + \ldots + x_7 = \sum_{i=1}^{7} x_i = 10 + 16 + \ldots + 19 = 98$ (as we found directly above). Sigma means 'the sum of'. When using sigma-notation, the total number of data is usually written as n or N. In order to find the mean we require the total to be divided by this number.

So in this example, write $n = 7$, $\sum_{i=1}^{n} x_i = 98$, and the mean $\bar{x}$ (x-bar) is then

$$\frac{1}{n}\sum_{i=1}^{n}x_i = \frac{98}{7} = 14.0.$$

This can be extended to frequency table calculations by using f_i to stand for the frequency of x_i. For the data of Example 1.2.2, the values x_i are 0, 1, 2, 3, 4, 5 and the corresponding f_i are 2, 3, 8, 7, 4, 1. The total now is the sum of each x_i times its corresponding f_i, $\Sigma_i f_i x_i$ (it is not always necessary to specify the full range of i when this is obvious from the context). So $\Sigma_i f_i x_i = (2 \times 0) + (3 \times 1) + \ldots + (1 \times 5) = 61$. The total number of observations is $\Sigma_i f_i$, so the mean is

$$\bar{x} = \frac{\Sigma_i f_i x_i}{\Sigma_i f_i}.$$

3.6 Measures of dispersion

A simple measure based on the two quartiles is the distance between them (the inter-quartile-range) or very often *half the distance from q to Q*, called the *semi-inter-quartile-range* (SIQR). This shows how scattered are the middle 50% of the data. When using the median as a measure of location it is natural to use the SIQR for dispersion. For Example 3.2.1, the original seven observations had $q = 10$ and $Q = 19$, giving SIQR $= \frac{1}{2}(Q - q) = \frac{1}{2}(19 - 10) = 4.5$. When the eighth observation is included, q becomes 11 and Q is 20·5, so that SIQR $= \frac{1}{2}(20·5 - 11) = 4·75$.

In Example 1.2.1, there is a middle pair, and $M = 47\frac{1}{2}$; half-way through the first 15 is the eighth observation in order of size so $q = 44$, and similarly $Q = 54$. Therefore SIQR $= \frac{1}{2}(54 - 44) = 5$. However, the quartiles have poor theoretical properties and cannot easily be tested or compared in statistical inference, although they are useful for display.

The usual measures of dispersion quoted are the *variance* or its square root, the *standard deviation*. To find the variance, the deviation of each observation from the mean is required, $(x_i - \bar{x})$ in the usual notation. The variance in a complete population of N observations is the average of the squares of these deviations,

$$\frac{(x_1 - \bar{x})^2 + (x_2 - \bar{x})^2 + \cdots + (x_N - \bar{x})^2}{N},$$

or in sigma-notation

$$\frac{1}{N}\sum_{i=1}^{N}(x_i - \bar{x})^2.$$

Almost always we have only a **sample**, not a complete population, and we must make an *estimate* of the population variance. For a random sample of n observations, the best estimate of variance is

$$\frac{1}{n-1}\sum_{i=1}^{n}(x_i - \bar{x})^2.$$

The divisor should be $(n-1)$, not n, in order to give an *unbiased* estimator, that is one which in regular use whenever the calculation is made will *on average* estimate the true variance.

The standard deviation of a set of data is expressed in the same units as the data themselves, whereas the variance is in squared units which makes for less easy understanding. However, the variance can be handled theoretically whereas the standard deviation cannot, and so the variance is used in statistical analysis and inference.

For Example 1.2.2, although the data were whole numbers the mean came to 2·44, so $(x_i - \bar{x})$ is always going to have two decimal places in it. If we do not carry all the figures when squaring $(x_i - \bar{x})$ we introduce a *rounding error*. There is an alternative way of calculating $\sum_{i=1}^{n}(x_i - \bar{x})^2$ which is usually easier. But first let us make the direct calculation. For $x_i = 0$, $(x_i - \bar{x}) = -2·44$ and $(x_i - \bar{x})^2 = 2·44^2 = 5·9536$. There were two zeroes, so the corresponding squared deviation is taken in to the sum twice. For $x_i = 1$ there were three observations and $(x_i - \bar{x}) = -1·44$, which when squared is 2·0736; this must be included three times. Proceeding in this way gives $(2 \times 5·9536)$ $+ (3 \times 2·0736) + (8 \times 0·1936) + (7 \times 0·3136) + (4 \times 2·4336) + (1 \times 6·5536) = 38·16$. For an unbiased estimate of variance this is divided by 24 to give 1·59. The standard deviation is the square root of this, namely 1·26.

There is an algebraic result which says that

$$\sum_{i=1}^{n}(x_i - \bar{x})^2 = \Sigma x_i^2 - \frac{1}{n}(\Sigma x_i)^2,$$

the sum of the squares of each observation *minus* the total squared and divided by n. Here

$$\sum x_i^2 = (2 \times 0^2) + (3 \times 1^2) + (8 \times 2^2) + (7 \times 3^2) + (4 \times 4^2) + (1 \times 5^2) =$$
$$0 + 3 + 32 + 63 + 64 + 25 = 187.$$

We know (from earlier calculation of the mean in Section 3.3) that $\Sigma x_i = 61$. Therefore

$$\sum_{i=1}^{n}(x_i - \bar{x})^2 = 187 - \frac{61^2}{25} = 187 - 148·84 = 38·16.$$

This arithmetic, using the original data, is much easier.

When data have been arranged in a frequency table, their sum is $\Sigma f_i x_i$, where i goes through the list of possible x-values. The sum of their squares is $\Sigma f_i x_i^2$. A complete table for calculating mean and variance is therefore:

x_i	0	1	2	3	4	5	Total
f_i	2	3	8	7	4	1	25
$f_i x_i$	0	3	16	21	16	5	61
$f_i x_i^2$	0	3	32	63	64	25	187

$$\bar{x} = \frac{\Sigma f_i x_i}{\Sigma f_i} = \frac{61}{25} = 2·44.$$

The estimated standard deviation is usually called s. So

$$s^2 = \frac{1}{n-1}\left[\Sigma f_i x_i^2 - \frac{(\Sigma f_i x_i)^2}{\Sigma f_i}\right] = \frac{1}{24}\left[187 - \frac{61^2}{25}\right] = \frac{38·16}{24} = 1·59, \text{ as above.}$$

Hence $s = +\sqrt{s^2} = 1 \cdot 26$.

The same type of calculation can be used for continuous data in a grouped frequency table, where the values x_i will be the mid-points of intervals and f_i the corresponding frequencies in the intervals. Of course this will not give exactly the same value as working from the 'raw' data because of the information we lose by grouping (Section 3.3).

For Example 1.2.1, the calculations from the raw data (the 30 observations) give $\Sigma x_i = 1507$ and $\Sigma x_i^2 = 77\,803$. The mean is $\bar{x} = \frac{1507}{30} = 50 \cdot 23$. The variance will be

$$s^2 = \frac{1}{29}\left[77\,803 - \frac{1507^2}{30}\right] = \frac{2101 \cdot 3667}{29} = 72 \cdot 4609.$$

Hence $s = +\sqrt{72 \cdot 4609} = 8 \cdot 51$.

When the data have been grouped into a frequency table with six intervals as in Section 3.3, these calculations can be repeated using the mid-points of the intervals as x_i. The table looks much better if set out in columns.

Interval	Mid-point x_i	Frequency f_i	$f_i x_i$	$f_i x_i^2$
38–43	40·5	6	243·0	9841·50
44–49	46·5	13	604·5	28109·25
50–55	52·5	4	210·0	11025·00
56–61	58·5	3	175·5	10266·75
62–67	64·5	2	129·0	8320·50
68–73	70·5	2	141·0	9940·50
Total		30	1503·0	77503·50

Now $\bar{x} = \frac{1503 \cdot 0}{30} = 50 \cdot 1$, almost the same as for the raw data; also

$$s^2 = \frac{1}{29}\left[77\,503 \cdot 50 - \frac{(1503 \cdot 0)^2}{30}\right] = \frac{2203 \cdot 20}{29} = 75 \cdot 9724, \quad \text{which gives } s = 8 \cdot 72.$$

If data are given in a table, with an 'open-ended' interval at one end (or both ends), it may not be at all obvious what to use for mid-points. The choice of mid-point has to be made using some intelligence, and any knowledge of the likely pattern of the data within an interval. For example, age-at-death data may contain an 'over 75 'class. We know people can live to be 100, so we might guess the interval as '76–100': in most populations this would leave out very few members indeed. But the mid-point is hardly likely to be 88 (or 87·5), since most of the individuals will die before the age of 88 and relatively few will live beyond that. That 'mid-point' should, for purposes of calculation, represent the way the data are spread through the interval, and if most of the data are likely to be nearer the lower end the mid-point should also be towards the lower end. In age-at-death data, 80 might be a realistic 'mid-point' for '76–100'.

3.6.1 Example
In a forestry nursery, the heights of trees of a *pinus* species after one season's growth are summarised in the following table:

Height (*m*)	Number of trees
Less than 0·2	18
0·2 and under 0·3	23
0.3 and under 0·4	30
0·4 and under 0·5	37
0·5 and under 0·6	35
0·6 and under 0·8	41
0·8 and over	28
Total	212

Using suitable mid-points for the intervals, calculate the mean and standard deviation of these data.

The first mid-point may be taken as 0·1, although that could be an underestimate, because if the trees survive at all they are likely to make a noticeable amount of growth; thus there may be many more between 0·1 and 0·2 than between 0 and 0·1. However, this is less of a problem than the choice of a mid-point for the top interval, '0·8 and over'. Without thinking, we might take 0·9; but we are not told what the maximum observed height was, and there were a considerable number of observations (28) in this upper interval. It seems likely that 0·9 will be too low. We will compare the results using, first, 0·9 and then 1·0 instead, as the mid-point for the upper interval. In each case we will take 0·1 as the mid-point for the lower interval.

Mid-point x_i	Frequency f_i	$f_i x_i$	$f_i x_i^2$
0·1	18	1·80	0·1800
0·25	23	5·75	1·4375
0·35	30	10·50	3·6750
0·45	37	16·65	7·4925
0·55	35	19·25	10·5875
0·7	41	28·70	20·0900
0·9	28	25·20	22·6800
Total	212	107·85	66·1425

With 0·9, $\Sigma f_i x_i = 107 \cdot 85$ giving

$$\bar{x} = \frac{107 \cdot 85}{212} = 0 \cdot 509$$

and $\Sigma f_i x_i^2 = 66 \cdot 1425$, so that

$$s^2 = \frac{1}{211} \left(66 \cdot 1425 - \frac{107 \cdot 85^2}{212} \right) = 0 \cdot 05\ 344$$

giving $s = 0 \cdot 231$.

If 1·0 is used, the entries for $f_i x_i$ and $f_i x_i^2$ in the last row of the table both become 28·0, so that now $\Sigma f_i x_i = 110 \cdot 65$, giving

$$\bar{x} = \frac{110 \cdot 65}{212} = 0 \cdot 522$$

and $\Sigma f_i x_i^2 = 71{\cdot}4625$, so that

$$s^2 = \frac{1}{211}\left(71{\cdot}4625 - \frac{110{\cdot}65^2}{212}\right) = 0{\cdot}06\,498$$

giving $s = 0{\cdot}255$.

The changes in the values of mean and standard deviation can hardly be called negligible.

3.7 When to use which measure

The choice between mean and median, for descriptive purposes, depends on whether data are roughly symmetrical or quite skew. The *mean* is a good summary of the 'centre' of a symmetrical distribution, and usually best for theoretical purposes such as significance tests. It is easily calculated (with the aid of a computer for large data sets) and is well understood as the 'average' size of all the data. Because it does use all the data values, it can be seriously affected by one or two values that are much larger or smaller than the others, even by one or two 'rogue' values that perhaps should not be included because they were not from the same population—but note that it is wrong to exclude them just because of size, unless there is (on further enquiry or study) some good biological reason for them. A tree in an orchard may grow much more slowly than others of the same variety, and on inspection there may be some damage to roots, perhaps from animals or from rocky soil. It would be wrong to discard the record for that tree without further inspection.

The *median*, as we have seen, is less affected by extreme values and so is often better for skew data. However, when theoretical work is to be done, e.g. significance testing, or confidence intervals, the mean is very often used: but we shall see later (Section 7.5) that the scale of measurement upon which records are taken may need to be altered by a 'transformation', after which the data are more nearly symmetrical.

The median and quartiles go together; so do the mean and standard deviation.

There is another 'central' measure, the *mode*. This is the true value having the highest frequency, or the interval containing the highest frequency. With small data sets it can be hard to decide where the mode is. It is rarely used.

3.8 MINITAB descriptions of data

Means, medians and standard deviations are sometimes called 'one-number' statistics: all the data are summarised by quoting just one number. Clearly we need more than one number in practice: we shall want one for location, one for dispersion, and possibly a choice between alternatives for each of these. The DESCRIBE command gives all the likely summary measures, and we can make a choice from them.

The total number, n, of observations is followed by the mean, the median, the standard deviation, the minimum and maximum observations, and the lower and upper quartiles (which are called Q1 and Q3). We may thus quote *either* the mean and the standard deviation *or* the median and the quartiles (and the semi-inter-quartile range); the minimum and maximum will be needed if we wish to draw a box-and-whisker plot (although Minitab also has this as an additional feature). DESCRIBE also gives the *standard error of the mean*, $s/\sqrt{n}$ (see Section 7.7), and one other measure which is less common. This other measure is called 'TRMEAN', and is the mean of the data *omitting* the smallest 5% and the largest 5% of the data after these have been arranged in rank order. The thinking behind this 'trimmed mean' is that if data are symmetrical then the two extremes which are omitted will roughly balance one another, and TRMEAN will have a very similar value to MEAN; this would not be so in a skew distribution.

3.9 Other measures found in some computer programs

A measure of *skewness* can be based on $\Sigma(x_i - \bar{x})^3$, the average of which is an estimate of the so-called 'third moment' of the data. It is zero for a symmetrical distribution; so if its value for a sample is small (positive or negative) the data are fairly symmetrical. This calculation is often included in descriptive packages written for industrial statistics, but we can in fact learn much more about the shape of a set of data from graphical plots. Apparent skewness, as calculated, can be caused by one or two strange or 'outlier' observations, and it is these we want to discover, not some overall measure.

Packages that contain this will also often have *kurtosis*, measured by $\Sigma(x_i - \bar{x})^4$. This shows whether a symmetrical distribution is sharply peaked in the centre, or relatively flat. Again we do not often want this measure for biological use.

Finally, there is the *mean absolute deviation*, measured by $\frac{1}{n}\sum_{i=1}^{n} |x_i - \bar{x}|$, where the modulus sign (vertical bars) means that the absolute size of the deviation of x_i from $\bar{x}$ is taken, ignoring the sign. The m.a.d. is the average of these deviations. It is another measure of dispersion, like the variance but simpler in arithmetic because there is no squaring. Unfortunately, the m.a.d. has no useful theory applying to it, so its value cannot be examined by a statistical test, and it is again rarely used. However, as with the standard deviation, if one set of data has a much larger m.a.d. than another, then the first set is much more scattered than the other. So the m.a.d. gives similar information to the s.d., but not in a form that can be used numerically.

3.10 Exercises on Chapter 3

3.10.1   The following yields (lb) were obtained from plots of a fixed size in a field of potatoes growing under the same fertiliser treatment. Calculate the mean, median, quartiles, variance and standard deviation of these yields.
28, 21, 14, 17, 24, 19, 22, 21, 16, 26, 20, 19, 23, 22, 20, 24, 21, 19,
17, 15, 18, 22, 23, 20, 21, 25, 22, 20, 18, 20, 22, 24, 26, 18, 24, 20.

3.10.2   Suppose that two additional plots in the same field as those listed in Exercise 3.10.1 gave yields of 29 and 30 lb. Examine how the measures calculated in 3.10.1 are changed by including these two plots in the data.

3.10.3   For the data on plant weights in Exercise 2.7.4, calculate the mean and standard deviation, the median and quartiles. Also draw a box-and-whisker plot for these data and comment on their symmetry (or lack of it).

3.10.4   For the data on plant heights in Exercise 2.7.5, calculate the median and quartiles directly from the data. Also use the cumulative frequency curve to estimate the median and quartiles.

3.10.5   Compare the two sets of data, from Sites A and B, given in Exercise 2.7.3, using suitable numerical summary measures. Comment on the results.

3.10.6   The weights of 12 hens' eggs are 48, 52, 47, 49, 53, 44, 53, 58, 57, 52, 59, 56. Calculate the mean and standard deviation of egg weight. We are now told that the first six were laid by hens of one breed, and the second six by hens of another breed. Calculate the mean and standard deviation for each breed, and comment.

Repeat, using the median and quartiles instead of the mean and standard deviation.

3.10.7   For the data quoted in Exercise 2.7.2, on the weights of animals, calculate the mean, median, variance and standard deviation; and comment on any difference found between the mean and median.

3.10.8   The total numbers of weeds of a particular species found in each of 50 sampling quadrats were 1, 1, 1, 1, 1, 15, 1, 6, 3, 1, 4, 1, 14, 1, 1, 4, and 0 34 times. Calculate the mean and variance and comment on the suitability of these as summary measures.

4 Populations, samples and inferences

4.1 Defining a population

The purpose in doing an experiment or conducting a survey is to discover information that will be of scientific use in future. It may be to discover what methods of cultivation, nutrition treatments, control of pests and diseases etc. will be needed for an important crop to grow better in a country or region; or it may be to develop a new drug as an improved treatment for a serious disease. In a laboratory experiment it may be the rate at which an organism grows under standard conditions.

The **population** is the whole collection of units of material, individual people, animals, plants etc. that could be included in the study. Thus the whole country or region in a crop trial is the population, but for scientific purposes it needs to be defined more precisely: specifically the population is all the farms, smallholdings or other units which grow (or could grow) the crop—a population of units of land. In a drug trial, all the people affected by, or at risk to, the disease make up the population; and in laboratory work we can go on repeating an experiment under controlled conditions and generate a population of results.

We remarked in Section 1.1 that for statistical purposes a population should consist of like units; any systematic variation between parts of a population should be removed before the natural variation can be studied statistically. Therefore in examining a very widespread population such as a whole region, it will often be necessary to split it into 'sub-populations', each of which is homogeneous within itself, and to study each separately before (if required) combining the results to cover the whole region.

Statistical measures like means, medians, standard deviations are designed to summarise *single homogeneous* populations, and it follows that there is no point in calculating them for a whole population when there are sub-populations into which it should have been split. Analysis of large sets of data by computer carries the danger of finding meaningless summaries unless some preliminary check, probably graphical, is made to confirm homogeneity. Obviously any biological knowledge about likely differences among sub-populations will be used as well.

Unless the population is defined carefully, we cannot properly say what our experimental results refer to. We may not know what needs to be measured either: if we are asked about the 'purity' of drinking water in an area, which of the various undesirable elements (in many countries these will have maximum permitted levels set by law) are we assessing purity by? The full list of measurements to be made on each unit must be written down before work begins; it is very often impossible to go back later to fill in gaps left by bad planning of the programme of work.

4.2 Samples and inferences

A *population* is almost always much too large to do experimental work on, or even to carry out a survey and make observations. A *sample* must be used instead: this is part of the population, often a very small part. If an experiment using, say, 50 plants or animals is going to be the basis for making statements or *inferences* about a very large population, those 50 plants must represent the whole population. More mistakes and false claims come from using badly collected data than from errors in the calculations made.

One serious mistake is to apply the results to conditions which were not studied in the experiment, or included in the survey. Work done in a laboratory, or even on a small scale in a greenhouse, on a crop, may not be reproducible in the field where conditions are not controlled in the same way. Results that apply to a crop in one region of a country may not apply in another region with a different climate. In trials of new drugs on patients at different hospitals, even quite small differences in the way the trial drugs are administered can lead to different results at each hospital. In order to keep natural variation to a minimum in a laboratory experiment we often try to control environmental conditions very closely; but we must realise that the results then apply only to those conditions. When a difference between the ways of treating material is established in such an experiment, there must be a follow-up in more general, 'normal', conditions before it is valid to claim a real difference.

Another danger is to carry out experimental work over a limited range of conditions and then expect the results to apply beyond this range. Examples of this often occur in linear regression (Chapter 14). We may establish a relation between temperature and the solubility of a chemical through an experiment using temperatures from 10° C to 60° C; this gives no evidence as to whether the relation (straight line, or curve, whatever it may be) still holds below 10° C or above 60° C. *Never generalise outside the range of available data!*

4.3 Random samples

Statistical theory requires that all the units of material used in collecting the data are independent members of the same population. If we have raised 100 trees of the same species in a nursery, all of which are suitable for planting out on to a permanent site, but we only require 30 of these for an experiment, 30 should be chosen by *random sampling*. Simple random sampling is a method in which all the available units in the population have the *same* chance (probability) of being selected to be a member of the sample. This is the only scientifically sound method of sampling, on the basis of which we can infer from our data to a wider population. However, it is rather tedious to organise, which is why other methods are suggested (see Section 4.5). An essential tool for random samples has always been a *table of random digits*, as given in Fisher and Yates' (1963) Tables[2] and many other books. A small run of random digits is provided as Table VI (page 204), and further remarks on 'randomness' follow in Section 4.4.

For the 100 trees, they should be numbered 01, 02, ... , 99, 00 (standing for 100) and a table of random numbers (random digits) used to supply a run of random pairs, such as the digits:

2 3 3 5 4 2 2 5 6 1 7 8 0 9 1 0 2 0 4 1 . . .

which yield the trees numbered

23, 35, 42, 25, 61, 78, 09, 10, 20, 41, . . . as the sample.

This must be done before any trees can be moved from the nursery, and a few spares may be kept in reserve in case any selected tree turns out to have some *serious* defect which makes it unrepresentative. It is wrong to reject a tree unless there is a real, clear scientific reason; it is certainly wrong to reject the largest or smallest just because of their size because that would give a serious underestimate of the natural variability.

When there are not too many in the population, we may put numbered slips of paper into a box, one for each population unit, and pick out slips one at a time, each after thoroughly shuffling them. Since we want a sample *without replacement* (we cannot use the same items twice), we should set aside each slip as it is selected and not return it to the box. To give each number the same chance of selection, all should be written on paper of exactly the same size and thickness. Clearly random sampling can take time!

It can also be wasteful. If there are 280 trees, we can number them 001–280; but then any run of three random digits from 281 upwards must be thrown away. To avoid this, let each tree be associated with more than one set of numbers: 281 is the same tree as 001, 282 is the same as 002, and so on up to 560 which is the same as 280. Further, 001 can be identified with 561, 002 with 562 and so on again through to 840 which is the same tree as 280. Waste of the remaining runs of digits, from 841 to 999 and 000, cannot be avoided because they are not sufficient to give another *complete* numbering of the population. Therefore to use them would give trees 001–160 (corresponding to 841–000) four chances of occurring to every three chances for trees 161–280; and this is not *simple* random sampling because the chances are not all equal.

If a population can be divided into recognisable sub-populations which are to be studied separately, then a random sample must be taken from each of these, and these samples must all be independent of one another: we *cannot* use the same set of numbers over again in each, but must obtain a new set for each sub-population.

4.4 Computer samples

Computers contain 'random number generators' and pocket calculators often have a RND key. Are they reliable? What properties should 'random' digits have? Every place in the sequence in a table is equally likely to be occupied by any of the digits 0, 1, 2, . . . , 9. Therefore, in the long run when large numbers of digits are taken:

(1) there should be about the same *frequency* of each digit;
(2) there should also be about the same frequency of each possible *pair* 00, 01, 02, . . . , 98, 99;
(3) the distance or *gap* between successive 0s should follow a geometric distribution with $p = 0.1$ (see Chapter 5); and
(4) the patterns in runs of four digits can also be studied (the so-called *poker test*).

Published tables of random digits such as those in Fisher and Yates' (1963) Tables[2] have had all or most of these tests applied to them; if they pass them they are considered 'random'. This really means we have found no evidence of systematic patterns in them.

Many earlier computers produced samples which failed one or more of these tests. The methods now developed to give satisfactory samples are quite complex and need considerable memory space. When 'random' numbers are obtained on a computer it is therefore necessary to check that they have properties like (1)–(4), and particularly to avoid sequences which cycle, that is give the same run of numbers several times if they are used on a large scale. This would lead to samples in which every population member did not have the same chance of appearing. A good RND routine should be supported by a statement in the operating manual saying which tests of randomness it has passed.

Because satisfactory computer methods need plenty of memory space, there remains the danger even now that smaller pocket-calculator programs will not give entirely acceptable results, and so they are best avoided in problems that require more than a very few random numbers.

4.5 Non-random sampling

Because random sampling takes time to organise, several other methods are used. *Judgement* sampling (sometimes unfortunately called *representative sampling*) is a subjective method: the experimenter looks at the rows of trees in the nursery (for example) and selects those which look as if they represent the whole population. This method almost always gives biased results, having the wrong average size and a bad estimate of the variability of size. It is easy to miss the small trees altogether, and the large ones may be ignored as 'unrepresentative'; each member of the population does not have the same chance of selection. The literature contains several examples of unexpected, subjective errors which have arisen when observers have tried to do their own sampling visually.

Systematic sampling would use every third tree (if a sample of one-third of all trees is wanted) along the rows. Possible errors here would be related to any trends running in phase with the sampling scheme, such as gaps left in rows for machinery to pass through. If there are no such trends, systematic sampling can be effectively random, and is much quicker; but we need to know enough about a population to justify a claim that there are no trends before this method is reliable.

If one tries to bypass the use of a table, and to prepare a random sample by first writing down a run of numbers 'haphazardly' out of the head, it is very unlikely that these numbers will satisfy the conditions that random digits ought to satisfy; in particular 0 will be followed by 0 less often than it should, likewise 1 by 1, 2 by 2 and so on—see Exercise 4.7.1.

4.6 Assumptions made in analysis

We shall find, when using statistical methods, that the theory behind them is based on simple random sampling, usually *with* replacement though a straightforward adjustment can be made in survey data when there is no replacement.

In a simple random sample, the *expected values* (average values when sampling is repeated) of sample means of the measurements taken are equal to the actual means in the population. The same is true for variances provided the divisor $(n-1)$, and not n, is used in the sample variance. The estimates of the population mean and variance obtained from random samples are called *unbiased* estimates. We shall prefer these, in this text, to other types of estimate that have been suggested.

4.7 Exercises on Chapter 4

4.7.1 Write down haphazardly a run of 200 digits. Count the numbers of 0s, 1s, . . . , 9s. Count also the number of times 0 is followed by 0, by 1, by 2, . . . etc., and do the same for each of the digits 0 to 9. Check whether the frequencies of the 10 digits are roughly equal. The results can be examined statistically using the chi-squared test (Chapter 12).

4.7.2 How would you use a table of random digits to select:
   (a) 15 members from a population of 750;
   (b) 10 members from a population of 250;
   (c) 10 members from a population of 300;
   (d) 20 sample units, each 1m square, in a rectangular field 25m × 40m in size?

4.7.3 A group of students in a college is to be sampled to find out their annual expenditure on books. There are 648 students altogether, 363 studying Arts subjects and the others studying Science subjects. Lists are available from each Faculty, Arts and Science, with students' names listed in alphabetical order in each list.

Select a sample of 40 students from this population, explaining carefully how this is to be done and justifying your particular choice of method.

4.7.4 The following practical methods have been suggested for selecting random samples. Consider whether each is likely to be satisfactory, and why; and if unsatisfactory propose a better method.

(a) A list of all adults living in a village is available, in the form of a voters' list for elections. In order to survey a random sample of households in the village, to see what type of domestic heating they have, names are picked at random from the voters' list.

(b) The doctor who serves this village has a card-index, arranged by families (one card for each family). In order to survey the state of the teeth of the children in the village, a random sample of cards is taken, and if there are children in the family shown on a selected card then a random choice is made from among these.

(c) A horticulturalist wants to take a 1-in-10 sample of the strawberry plants growing in rows in a field. To save time selecting several random numbers, he first chooses a starting point at random among the first 10 plants in the first row. After this he takes every tenth plant down the first row, back along the second row, down the third row and so on through the field.

4.7.5 If a calculator with a RND function is available, obtain as many random numbers as time permits and repeat the analysis of Exercise 4.7.1. Keep the results for further study after the chapter on the chi-squared test (Chapter 12).

5 Probability, the binomial and geometric distributions

5.1 Probability

Statistical methods depend on **probability** theory, which is an important branch of mathematics. There is no need for us to study the theory here provided we are always careful, when using statistical methods, to check the list of conditions under which a method works satisfactorily. A practical working definition of probability will be sufficient. For a full development of probability theory at an appropriate level, see Clarke & Cooke (1992)[3].

The simplest way to understand probability is to use the idea of *proportional frequency*. Suppose that a population of items contains 50 with a particular genetic characteristic C (for example, a plant with red flowers) and 200 without it (not–C: for example, white flowers). The proportional frequency of that characteristic in the whole population is $\frac{50}{250}$ (or $\frac{1}{5}$). In simple random sampling, every item has the same probability (or chance) of selection. We are interested in the probability that we shall actually select a C when we choose one item at random from the whole population. We define this probability to be equal to $\frac{1}{5}$, the proportional frequency of Cs in the population.

Probability of selection of C at a single random sampling = proportional frequency of C in the population at the time of selection.

Suppose we select a second member. If we have *replaced* the first before we make the selection, we still have 50 Cs and 200 not–Cs, and the probability in the second selection is the same as the first. But if we have not replaced the first one, then the second sample *without replacement* has a different result depending on which type was selected first. If this was a C, we have 49 Cs left, and 200 not–Cs, so that selecting a C now has probability $\frac{49}{249}$. When the first was not a C, the second will be C, in sampling without replacement, with probability $\frac{50}{249}$.

Studying repeated samples from natural populations is much easier if we can assume that sampling has been made *with replacement*. In practice this is often not possible; but even then the population from which we might have sampled may be so large that we can effectively ignore the very small difference that removing one member would make. For example, a genetic study will assume that the population could go on reproducing according to the same rules indefinitely, and so the probability of the type we are interested in remains the same. If this is a reasonable assumption it makes the analysis of results much simpler and allows well-known methods to be used.

When probabilities remain the same at successive samples (or *trials*—a common technical-jargon name for samples), the result of the first sampling gives no information at all about the second. The two samples are *independent* statistically. This also simplifies analysis.

In both situations, sampling *with* and *without replacement*, we can follow through the possible sequences of Cs and not–Cs when several selections are made. First we require some end-points for the probability scale. Suppose there were 0 Cs and 250 not–Cs in the population. Then the probability of selecting a C is $\frac{0}{250} = 0$: this is an *impossible* happening (or *event*). At the other extreme, if there were 250 Cs and nothing else, the probability of selecting a C would be $\frac{250}{250} = 1$: this is a *certain* event. All probabilities must lie in the range from 0 to 1.

5.2 Tree diagrams

A useful method of discovering the probabilities of all the possible results (or *outcomes*) of repeated sampling is the *tree diagram*. It illustrates the idea very well for two or three samples but becomes too cumbersome for more: however, it points to the appropriate calculations which lead to the *binomial distribution*.

5.2.1 Example

Suppose three items are selected from the same type of population as described in Section 5.1 above, which contains $\frac{1}{5}$ of type C and $\frac{4}{5}$ of not–C; but now we shall assume it is so large that removing individuals from it does *not* alter these probabilities (effectively sampling with replacement or sampling from an 'infinite' population). We also assume successive samples are independent. In that case, using the idea of proportional frequency, $\frac{1}{5}$ of the first selections will be C and $\frac{4}{5}$ not–C; following this selection, whichever type it was, will be a C in $\frac{1}{5}$ of cases and a not–C in $\frac{4}{5}$ of cases. The third selection follows the same rules and there are eight possible orders in which the process could go:

$$C\,C\,C; \; C\,C\,\text{not} - C; \; C\,\text{not} - C\,C; \; \text{not} - C\,C\,C; \; \cdots; \; \text{not} - C\,\text{not} - C\,\text{not} - C.$$

The proportion of samplings of the first sort will be $\frac{1}{5} \times \frac{1}{5} \times \frac{1}{5}$; the second sort $\frac{1}{5} \times \frac{1}{5} \times \frac{4}{5}$; ... and the final sort $\frac{4}{5} \times \frac{4}{5} \times \frac{4}{5}$. We *multiply* probabilities (proportions) to find the probability of each of the possible orders: multiply when selections are made one *after* the other.

The tree diagram gives us all the possible outcomes and their probabilities. Only one of these outcomes (the different orders) can happen in any particular sampling: when one occurs, none of the others can. These outcomes are *mutually exclusive*. They are also *exhaustive* in the sense that between them they account for all possible outcomes. Therefore one, and only one, of them must happen. *The sum of the probabilities of a set of mutually exclusive, exhaustive outcomes is 1.*

Very often, the basic record of the order in which results occurred is not what we need; the real interest is in how many Cs there were among those sampled. In order to find probabilities for 0, 1, 2, or 3 Cs, which are the only possibilities, we must combine the different orders that give rise to one C or two Cs; there is only one order that gives all three Cs and only one that gives no Cs. When we combine orders like this we must *add* their probabilities: add when we are collecting together all the ways in which the same result (e.g. two Cs) can happen.

There are no Cs if we observe not–C not–C not–C, which has probability $(\frac{4}{5})^3 = (0\cdot8)^3 = 0\cdot512$.

One C may arise from three different orders, each with probability $(\frac{4}{5})^2(\frac{1}{5}) = (0\cdot8)^2(0\cdot2) = 0\cdot128$; the total probability of one C is then $3 \times 0\cdot128 = 0\cdot384$.

Similarly two *C*s arise with probability $3(\frac{4}{5})(\frac{1}{5})^2 = 0.096$; and three *C*s will have probability $(\frac{1}{5})^3 = (0.2)^3 = 0.008$. Setting these results out in a table:

Number of *Cs*	0	1	2	3	Total
Probability	0.512	0.384	0.096	0.008	1.000

Figure 5.1 illustrates how all the possible results are worked out.

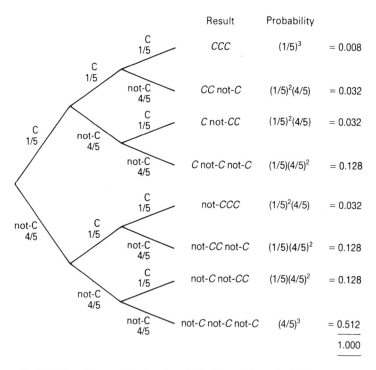

Fig. 5.1   Tree-diagram showing the calculations of Example 5.2.1.

5.3 Sampling without replacement

If we have a closed population, say of 40 apples in a box, and they are of two sorts, say skin-damaged and clean, a quality control process may involve taking a random sample of two of the apples and examining them. Now the first one will *not* be returned before the second selection is made; we require two distinct items for the sample. The situation may again be illustrated by a tree diagram, as shown in Fig. 5.2.

If we know the number of damaged apples in the box, we can assign probabilities for each stage of the selection. Suppose there are eight with skin damage; then the figures shown in the diagram will describe the process. The probabilities add up to 1.

The summary in terms of number damaged is:

Number damaged	0	1	2	Total
Probability	0.6359	0.3282	0.0359	1.0000

FIRST SAMPLE   SECOND SAMPLE   NUMBER DAMAGED   PROBABILITY

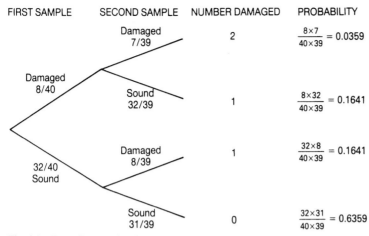

Fig. 5.2 Tree-diagram showing probabilities of damaged fruit.

Sampling from a closed, *finite* population therefore gives different results from those for a large 'infinite' population, the difference being more noticeable the smaller the population is.

Care must be taken in any practical problem to decide which type of sampling is appropriate.

5.4 Distributions

A *distribution* is a list of all the possible results in the sampling, together with a list of the probabilities corresponding to each result. The examples 'number of Cs' (in Section 5.2) and 'number damaged' (in Section 5.3) are both distributions. There are some standard distributions which arise when certain conditions are satisfied and we now consider some that are useful in biology.

In Section 5.5 we examine the *binomial* distribution, which is very useful for discrete data under the right conditions, and often arises in biological applications. A change in some of the conditions leads instead to the *geometric* distribution (Section 5.6), and another very important discrete distribution in biology is the *Poisson*, which will be studied in Chapter 6. These three distributions are the most useful for discrete data; and then in Chapter 7 we consider continuous data, and especially the *Normal* distribution.

5.5 The binomial distribution

This applies when a population consists of two types of member, as in the genetic example 5.2.1, and it is large enough to assume that (i) samples are independent of one another and (ii) the probability of the special type C does not alter. We take a random sample of n items from the population and wish to study the distribution of the number of Cs in this sample, which we call r.

There is only one way in which the result 'no Cs' can happen: every selection has to be a not–C. But there are n ways in which one C can arise: it can be in any one of the n possible places in the order of selection. When we consider finding two Cs we have to look for all possible pairs of places in the order, and so on. This rapidly becomes very hard to appreciate, but there is an arithmetical expression which calculates the number of orders in which we can arrange r Cs and $n-r$ not–Cs.

To simplify writing down this expression, *define* 'n-factorial', which is written $n!$, as $n(n-1)(n-2)\ldots 2.1$, the product of all integers from n down to 1. Then the required expression is

$$\frac{n!}{r!(n-r)!}, \text{ for } r = 1, 2, 3 \text{ to } (n-1);$$

when $r=0$ or n the expression is taken to be 1.

Examples of this calculation are as follows. For $n=10$, $r=1$,

$$\frac{n!}{r!(n-r)!} = \frac{10!}{1!9!} = \frac{10.9.8.7.6.5.4.3.2.1}{1.9.8.7.6.5.4.3.2.1} = 10;$$

there are 10 different ways of ordering one C and 9 not–Cs (as we noted above).

For $n=8$, $r=3$,

$$\frac{n!}{r!(n-r)!} = \frac{8!}{3!5!} = \frac{8.7.6.5.4.3.2.1}{3.2.1.5.4.3.2.1} = 56;$$

these expressions usually simplify considerably by cancelling out common factors from numerator and denominator.

The *conditions for a binomial distribution* to arise are as follows:

(1)　sampling is from an infinite population, containing two types of member;
(2)　the type being studied, C, exists in the population in a proportion p of members; that is to say, the probability of a C at each random selection is p;
(3)　each selection of a sample item is independent of all others;
(4)　a fixed sample size, n, is set;
(5)　the record made is the number, r, of Cs in the sample of n members.

Then, over repeated samplings, r will follow the binomial distribution in which the probability of r, which is usually written $P(r)$, is given by the formula

$$P(r) = \underset{\substack{\text{no. of ways of} \\ \text{splitting } (r, n-r)}}{\frac{n!}{r!(n-r)!}} \times \underset{\substack{\text{probability of} \\ \text{any one of them}}}{p^r(1-p)^{n-r}}$$

for $r = 0, 1, 2, \ldots, (n-1), n$.

The name 'binomial' is appropriate because there are two types in the population, and also the expression in factorials comes from the successive coefficients in the expansion of a binomial series.

The important feature in a binomial population is not usually its location or dispersion but the value of p. We shall develop methods of studying this (though they will in fact use the mean and standard deviation indirectly).

The mean in a binomial distribution is np and the variance is $np(1-p)$. If a set of data are binomially distributed, but p is not known, we estimate p by finding the mean of the data, $\bar{r}$, and dividing that by n. This uses a common method of estimating parameters: we are finding p by making the statement *Sample Mean = Population Mean*. Since we know n (by fixing the size of samples to be studied), this immediately gives an estimate of p.

5.5.1 Example
If $p = \frac{1}{4}$ and $n = 5$, the six possible values of r are 0, 1, 2, 3, 4, 5. Using the binomial formula,

$$P(r = 0) = 1(\tfrac{1}{4})^0(\tfrac{3}{4})^5 = 0{\cdot}2373$$
$$P(r = 1) = 5(\tfrac{1}{4})^1(\tfrac{3}{4})^4 = 0{\cdot}3955$$
$$P(r = 2) = 10(\tfrac{1}{4})^2(\tfrac{3}{4})^3 = 0{\cdot}2637$$
$$P(r = 3) = 10(\tfrac{1}{4})^3(\tfrac{3}{4})^2 = 0{\cdot}0879$$
$$P(r = 4) = 5(\tfrac{1}{4})^4(\tfrac{3}{4})^1 = 0{\cdot}0146$$
$$P(r = 5) = 1(\tfrac{1}{4})^5(\tfrac{3}{4})^0 = 0{\cdot}0010$$

These probabilities sum to 1, as they should. Since $p = \frac{1}{4}$, it is the smaller values of r which are most likely to occur.

Given the probability distribution, as here, we can calculate probabilities of composite events such as:

(i)   there are more than two of the special type;
(ii) there is at least one of the special type.

For (i), $r = 3$ or 4 or 5 and these three probabilities must be added to give the answer 0.1035. We can interpret this to say that only about 10% of samples, in the long run, are expected to give more than two members of special type. For (ii) note that *either* there is at least one of special type *or* there are none; the probabilities of these two mutually exclusive, exhaustive events must add to 1. So the required answer is $1 - P(r = 0) = 1 - 0{\cdot}2373 = 0{\cdot}7627$. The same answer would be found by adding together $P(r = 1)$, ... up to $P(r = 5)$, but is more quickly found as $1 - P(r = 0)$. Whenever the words *at least* appear in the specification of an event, it is worth considering whether to calculate the probability directly or by an indirect method such as this.

5.5.2 Example
When $p = \frac{1}{2}$, the calculation is very simple, and produces a symmetrical pattern of distribution as shown in Fig. 5.3(a) for $n = 4$. The formula gives

$$P(r) = \frac{4!}{r!(4 - r)!}\left(\frac{1}{2}\right)^r\left(\frac{1}{2}\right)^{4-r} \qquad \text{for } r = 0,\ 1,\ 2,\ 3,\ 4.$$

The first part of this product, involving the factorials, has the value 1 when $r = 0$ or 4; 4 when $r = 1$ or 3; and 6 when $r = 2$. The second part, $\left(\frac{1}{2}\right)^r\left(\frac{1}{2}\right)^{4-r}$, is equal to $\left(\frac{1}{2}\right)^4$ which is $\frac{1}{16}$, for all the values of r. Therefore

$$P(0) = \frac{1}{16} = P(4)$$
$$P(1) = \frac{4}{16} = \frac{1}{4} = P(3)$$
$$P(2) = \frac{6}{16} = \frac{3}{8}.$$

The sum of these five probabilities is 1, as it must be since they account for all the possible values of r. This distribution would be a statistical model for the number of male children in families of four children altogether, *provided* that the probabilities of male and female births were both equal (so that $p = \frac{1}{2}$) and that the sexes of births of individual children were independent of one another; in practice in the human population neither of these conditions may be exactly true, and the probability of a male birth is very slightly above $\frac{1}{2}$.

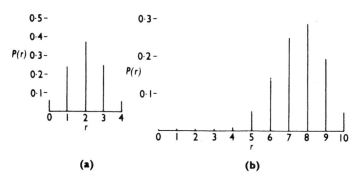

(a)   **(b)**

Fig. 5.3 Bar-diagrams for the values of $P(r)$ in the binomial distributions (a) with $n = 4$, $p = \frac{1}{2}$ (Example 5.5.2); (b) with $n = 10$, $p = \frac{3}{4}$ (Example 5.5.3).

5.5.3 Example

Consider Fig. 5.3 (b), illustrating the binomial distribution with $n = 10$ and $p = \frac{3}{4}$. If we study the results of a simple genetic trial where a 3:1 segregation of types $A:a$ is expected, the proportion of As is $p = \frac{3}{4}$. The number, r, of As in random samples from 10 members of the population should then follow this distribution.

The possible values of r are 0, 1, . . . , 9, 10 and the binomial formula gives

$$P(r) = \frac{10!}{r!(10-r)!} \left(\frac{3}{4}\right)^r \left(\frac{1}{4}\right)^{10-r} \qquad \text{for } r = 0, 1, \cdots, 9, 10.$$

The values of $P(r)$ are found to be

$r = 0$	$P(r) = 0.000\ 001$	$r = 6$	$P(r) = 0.145\ 998$
1	0.000 029	7	0.250 282
2	0.000 386	8	0.281 568
3	0.003 090	9	0.187 712
4	0.016 222	10	0.056 314
5	0.058 399		

Again they must add to 1 (give or take a very small rounding error, in this case 1 in the last place of decimals).

The result *no As* has probability 1 in one million; we may interpret this by saying that it is extremely unlikely that one random sample of 10 members will contain no *As*. The maximum probability is for $r=8$, and because $p=\frac{3}{4}$ we generally expect to find a considerable majority of *As*. In fact the probability of *not more than four As*, which is $P(0)+P(1)+P(2)+P(3)+P(4)$, is 0·019 728, or roughly 0·02; only 2% of random samples of 10 members will contain four or less *As*. So again this event would be very unlikely if we were studying only one sample.

The mean value of r is $np=10\times\frac{3}{4}=7\cdot5$, and this is the average number of *As* we would expect to find per sample if we studied a large number of random samples of size 10. The variance of r from sample to sample is $np(1-p)=10\times\frac{3}{4}\times\frac{1}{4}=1\cdot875$.

5.5.4 Example. Comparing a sample with a hypothesis

Six plants have been measured on each of 40 experimental plots. The numbers of records ending in 0 or 5 are noted for each plot, the data being measurements (mm) supposedly to the nearest whole number. So if there is no recording bias towards 0 and 5, the proportion of 0s and 5s should be $p=0.2$. Samples of $n=6$ have been taken, so the number of 0s and 5s in each sample should follow a binomial distribution with $n=6$ and $p=0\cdot2$. The binomial probabilities will therefore be calculated from the formula:

$$P(r) = \frac{6!}{r!(6-r)!}(0\cdot2)^r(0\cdot8)^{6-r} \qquad \text{for } r = 0,\ 1,\ 2,\ 3,\ 4,\ 5,\ 6.$$

The probabilities when multiplied by 40 give the expected distribution of values of $r=0, 1, 2, 3, 4, 5, 6$ in the 40 samples. The following table gives the actual data, on the first two rows, then the binomial probabilities and finally the expected frequencies in 40 samples.

$r=$	0	1	2	3	4	5	6	Total
$f=$	6	12	14	5	1	1	1	40
$P(r)=$	0·2621	0·3932	0·2458	0·0819	0·0154	0·0015	0·0001	
$\times 40=$	10·48	15·73	9·83	3·28	0·62	0·06	0·00	

The second and fourth rows do not agree very well, and we need methods of comparing these so-called *observed* and *expected* frequencies (see Chapter 12).

The *mean* of a binomial distribution is np, and in Example 5.5.4 it should therefore be $6\times0\cdot2=1\cdot2$. From the observed data, the mean $\bar{r}$ is

$$\frac{(6\times0)+(12\times1)+(14\times2)+(5\times3)+(1\times4)+(1\times5)+(1\times6)}{40}=\frac{70}{40}=1\cdot75,$$

which is considerably more than 1·2 and therefore suggests that there may be too many 0s and 5s.

The *variance* of a binomial distribution is $np\,(1-p)$. If the observed data do not appear to have a variance of about this value, one of the conditions listed in Section 5.5 may have broken down. Either the sampled items may not be independent of one another or the value of p may be varying as sampling continues. The variance in this set of data requires Σfr^2 to be calculated also: this is 190. Hence

$$s^2 = \frac{1}{39}\left(190 - \frac{70^2}{40}\right) = 1\cdot 73,$$

which compares very poorly with the theoretical value of $6 \times 0\cdot 2 \times 0\cdot 8 = 0\cdot 96$. If the binomial conditions do not really apply, then of course an 'estimate' of p is of no use. For example, when there is no reason for p to remain constant from one member, or one part, of the population to another, the binomial will *not* give a satisfactory explanation of observations; for the binomial to be applicable, it is essential that every member sampled shall be subject to the *same* probability or proportion p. We must always remember that it is extremely important to make sure that the population being studied is *homogeneous* before applying statistical methods to it. When studying genetic characteristics using the binomial, what matters is that every member examined shall have the *same* probability of showing the particular characteristic being studied.

5.6 The geometric distribution

Another form of sampling in a population containing two types is not to fix the sample size n but instead to continue sampling until one of the special type C is found, and to count the sample size that was necessary to achieve this.

The *conditions for a geometric distribution* are

(1) as for the binomial (Section 5.5);
(2) as for the binomial (Section 5.5);
(3) as for the binomial (Section 5.5);
(4) counting continues until a member of special type is found;
(5) the record made is of the sample size, n, which was necessary to find the first C.

Then, over repeated samplings, n will follow the geometric distribution in which the probability of n is given by the formula

$$P(n) = p(1 - p)^{n-1} \qquad \text{for } n = 1,\ 2,\ 3,\ldots.$$

This formula arises because *all* the first $(n-1)$ sampled items must be 'not–C', and so there are $(n-1)$ factors $(1-p)$; then the last one is a C, which gives one factor p. There is of course only one order in which this result occurs: the C has to be the last one of the n items sampled. So nothing like a binomial coefficient comes into the formula; and the probabilities have to be multiplied together because items are being sampled one after the other. There is no theoretical upper limit to the possible values of n: we do not know how long it will take to find the first C.

5.6.1 Example
Count the number of other genetic types found in a sample up to the first double dominant. Repeat this process 60 times ($r = 1$ means a double dominant came first.)

r:	1	2	3	4	5	6	7 or more	Total
f:	38	16	4	0	1	1	0	60

The distribution is called 'geometric' because the successive probabilities are the terms in a geometric progression with the first term p and common factor $(1-p)$. The mean is

$1/p$ and the variance is $(1-p)/p^2$. The value of p in a population can be estimated by calculating the mean $\bar{r}$ of a random sample from the population and then using '*Sample Mean = Population Mean*' to give $\bar{r}=1/\hat{p}$, so that the estimate $\hat{p}$ of p is $1/\bar{r}$. For Example 5.6.1, $\bar{r}$ is $\frac{93}{60}=1.55$, and so $\hat{p}=\frac{1}{1.55}=0.645$. This is the estimate of the probability of a double-dominant at each selection from the population.

Sometimes geometric sampling or its extension (see Section 5.7), 'inverse binomial sampling', can be a good way to estimate the value of p in a population, when p is very small (near zero). It can be less wasteful of sampling time than setting a fixed sample size n and finding very few Cs, i.e. a very small value of r in a relatively large n.

5.7 The negative binomial distribution

The conditions for this are exactly like those for the geometric distribution except that n is the number required up to and including the kth of special type. The value of k is nominated beforehand, and then n may take any value from k upwards without any upper limit.

$$P(n) = \frac{(n-1)!}{(k-1)!(n-k)!}\, p^k(1-p)^{n-k}, \qquad n = k,\ k+1,\ k+2,\dots.$$

(These are successive terms in a binomial expansion with negative index; hence the name.)

The mean of this distribution is k/p and its variance is $k(1-p)/p^2$.

There are other sets of conditions which can also lead to a negative binomial distribution, and these will be seen after the Poisson distribution has been studied (Chapter 6).

5.8 MINITAB computing

MINITAB will provide a print-out of all the values of $P(r)$ in a binomial distribution, for $r = 0, 1, 2, \dots, n$ once n and p have been specified; it will also give any individual probabilities that may be asked for. (It omits from the list any very small probabilities, those less than 0.000 05.)

The *cumulative probabilities* can also be obtained: these are the total probabilities that r is *less than or equal to* a given value, e.g. the cumulative probability for $r = 2$ is $P(0) + P(1) + P(2)$. These may be useful in their own right, or in finding probabilities of events such as 'r is at least 3', which would be $1 - P(0) - P(1) - P(2)$.

Another piece of information which will be useful later on is to find the value of r which has some specified cumulative probability, for example 0.9. In most cases (because we are dealing with a discrete distribution) the output will consist of two values of r, that with cumulative probability just below 0.9 and that with cumulative probability just above.

5.9 Exercises on Chapter 5

5.9.1   If the four types *AB*, *Ab*, *aB*, *ab* occur in the ratio 9:3:3:1 in a population, what is the probability that

(a) one randomly selected member will be an *AB*?
(b) one randomly selected member will be either an *AB* or an *Ab*?
(c) when two members are selected at random, the first will be *aB* and the second *Ab*?
(d) when two members are selected at random, one will be *AB* and the other *ab*?

5.9.2   A plant breeder knows by experience that when he crosses two varieties of currant, 10% of the offspring will be disease-resistant, and that only 20% of the offspring will be sufficiently vigorous growers to warrant further trial. If he wishes to breed plants which are both vigorous and disease-resistant, and these two characteristics are independent of one another, what proportion of offspring will be of use to him?

What is the probability that, in 100 such offspring, none will be worthy of further study? And what is the mean value of the number of useful offspring in 100?

5.9.3   Seeds of a certain plant have a 60% germination rate. Calculate the probability that, when eight of these seeds are planted, six or more will germinate.

5.9.4   Persons suffering from a blood disease are found to have an abnormality in one particular chromosome. However, not all samples of this chromosome are abnormal, and in order to estimate the proportion of affected ones, five examples of this chromosome are examined from each of 120 patients, the number of affected ones, *r*, being recorded for each patient. The results are tabulated below; estimate the proportion *p* of affected chromosomes.

r:	0	1	2	3	4	5	Total
f:	6	31	42	29	10	2	120

5.9.5   Explain why you would, or would not, expect the measurements described below to follow a binomial distribution. For those which do, state the values of *n* and *p*.

(a) Twelve animals are selected at random from a large population in which a particular type of marking occurs with probability $\frac{1}{3}$ on each member. The number having the marking in the sample is recorded.

(b) A penny is tossed until it first shows a head. The number of tosses required for this is recorded.

(c) A group of 40 students contains 10 who are left-handed. Five students are chosen at random from this group, and the number of left-handed members in these five is recorded.

(d) An animal is being trained to traverse a maze correctly. At each session it is allowed five runs at the maze, and the number of correct runs is recorded.

(e) Groups of four people each answer a psychological test in the form of a battery of questions, and it is expected that any person has an equal chance of scoring above or below half the total 'marks'. The number in each group scoring above half is recorded.

5.9.6 In a food production process, packaged items are sampled as they come off a production line: a random sample of five items from each production batch is checked to see if each is tightly packed. A batch is accepted if all five sample items are satisfactory, and rejected if there are three or more unsatisfactory packages in it; otherwise a further sample is taken before making a decision. If in fact the packing machine is giving 80% of items properly packed, what is the probability that this second sample will be necessary?

The second sample also consists of five items. What is the probability that, out of the two samples (10 items), there are nine satisfactory?

What are the assumptions behind your calculations?

5.9.7 (a) Assuming that there are equal numbers of male and female births, what is the probability that a family of six children will contain four boys and two girls?

(b) What is the probability that the deviation from equality of sexes in the family will be as large as that in (a) *or larger*?

5.9.8 Two apparently normal parents have a child which suffers from phenylketonuria, a disease which is due to a recessive gene. On testing, the parents are found to be carriers of the gene, and it is therefore to be expected that $\frac{1}{4}$ of their children will be affected. If they have two more children, what is the probability that both will be normal?

5.9.9 A fruit grower is assessing the proportion of Grade I apples in his crop by watching as they pass over a mechanical grader. Each time a Grade I apple is found, he begins counting again from zero, until the next Grade I appears. (He includes the Grade I apple in his count.)

Fifty counts were 1, 7, 2, 3, 2, 4, 10, 1, 4, 5, 3, 3, 6, 2, 2, 4, 1, 3, 7, 6, 4, 8, 3, 3, 9, 5, 1, 1, 3, 2, 12, 8, 3, 2, 3, 4, 1, 5, 3, 1, 2, 2, 4, 6, 2, 3, 2, 1, 11, 4.

Estimate the proportion of Grade I apples in his crop. What assumptions must be made for this estimation process to be valid?

6 The Poisson distribution

6.1 Limiting form of the binomial

This distribution was first discovered by the mathematician, S. D. Poisson, as a 'limiting form' of the binomial distribution when n is very large and p is very small. In such a case the probabilities for all except the first few values of r will be very small and it would not be worth computing them; and also the factorial expressions lead to a lot of arithmetic. Poisson wrote $\mu = np$, the mean of the distribution, and showed that by using the exponential function the values of $P(r)$ may be computed more simply as

$$P(r) = \frac{e^{-\mu}\mu^r}{r!} \text{ for } r = 0, 1, 2, \ldots.$$

Again, as with the geometric and negative binomial distributions (Sections 5.6 and 5.7), there is no theoretical upper limit to possible values of r.

The variance, as well as the mean, is equal to μ. The original derivation gave the Poisson the title of the distribution for *rare events*. While this is true, it is not the most important reason for the Poisson to occur in real data. There is a set of conditions which may often be satisfied by various different types of data.

6.2 Conditions for the Poisson distribution to arise in practice

These conditions may be summarised in the words *randomness, independence, constant rate*. In more detail, they are:

(1) events occur *at random* in continuous space or time;
(2) events occur singly;
(3) events occur uniformly, i.e. at *constant rate*;
(4) events occur independently, i.e. the occurrence of one event has no influence on the next;
(5) the number, r, of events in unit space or time will then follow the Poisson distribution.

When conditions (2) and (4) break down, and the events occur in groups or clumps, not singly and independently, the negative binomial (Section 5.7) may fit data better (provided the *groups* arise at random).

To illustrate these conditions, consider counting small insects on the leaves of a host plant, and suppose the insects do not tend to group together nor do they show any preference for any particular position (upper, lower) or age (older, younger) of leaves. The number of insects per unit area is r, and counting is done on several unit areas of the leaves of several plants.

Condition (3) above implies that any particular spot in the area is equally likely to have an insect on it; and conditions (1) and (4) imply that presence of an insect at one particular spot on a leaf has no bearing on whether or not another insect will be very close by. If the insects had emerged from a batch of eggs laid on the leaves, these conditions, as well as condition (2), would be likely not to hold.

The actual unit of area (in this case—or of time, or volume, as appropriate) can be whatever is practically convenient: it does not need to be 1 cm^2, 1 second, 1 ml provided it is constant for all the observations made. If conditions (1)–(4) do apply, they apply whatever we choose as unit.

6.2.1 Example

Suppose that we count the number, r, of a particular type of cell present in unit area of a suspension on a microscope slide, and find that, on average, there is one of these cells in every two units of area, i.e. $\mu = \frac{1}{2}$. If the counts r follow a Poisson distribution, we shall want to calculate the probabilities given in the formula

$$P(r) = \frac{e^{-\mu}\mu^r}{r!} \qquad \text{for } r = 0, 1, 2, \dots.$$

The number e is a mathematical constant (the base of natural logarithms) and is equal to 2.718 28. . . ; this is built in to most pocket calculators and computers, sometimes called EXP—short for *exponential*. The expression $e^{-\mu}$ means $1/e^\mu$. Hence

$$P(0) = e^{-\mu} = e^{-\frac{1}{2}} = \frac{1}{e^{\frac{1}{2}}} = \frac{1}{\sqrt{e}} = \frac{1}{\sqrt{2 \cdot 718\,28}} = 0 \cdot 6065.$$

Also

$$P(1) = \frac{e^{-\mu}\mu^1}{1!} = \frac{1}{2}e^{-\frac{1}{2}} = 0.3033; \quad P(2) = \frac{e^{-\mu}\mu^2}{2!} = \frac{(e^{-\frac{1}{2}})(\frac{1}{2})^2}{2 \times 1} = 0 \cdot 0758;$$

and similarly $P(3) = 0.0126$, so that $P(r = 4 \text{ or more})$, written $P(r \geqslant 4)$, is 0.0018.

The sum of all the probabilities $P(r)$, taken over all possible values of $r = 0, 1, 2, 3$ . . . without upper limit, must equal 1; in fact the first few values $P(0)$, $P(1)$, $P(2)$, $P(3)$ add up to almost 1 in this example, and so we express the probability of all the remaining values, $P(r \geqslant 4)$, in a single number. The Poisson distribution tails off very quickly when μ is small, and quite often it is necessary to work out only relatively few probabilities.

6.2.2 Example

Figure 6.1 illustrates the following set of data, in which r denotes the number of radioactive particles emitted in unit time (which here was $\frac{1}{2}$ minute) by specimens of plant material grown in a solution containing a labelled nutrient. Altogether 370 separate half-minute periods were observed, and the summary of the counts is given as a table:

Number of particles, r:	0	1	2	3	4	5	6	7	8
Frequency f:	2	21	37	50	79	66	49	36	19

Number of particles, r:	9	10	11	12	13	14	15	16+	Total
Frequency f:	7	2	0	0	1	0	1	0	370

This is a typical pattern for the Poisson distribution when μ is greater than 1, the maximum frequency being reached quickly (here it is at $r=4$) and the bar-diagram tailing off to the right more slowly. The mean of this set of data will be found to be 4.59 and the variance 4.41. We shall study these data again later, and find that the Poisson distribution is a satisfactory model for them. We may therefore feel justified in suggesting that the three conditions of randomness, independence and constant rate held during the period while these data were being collected.

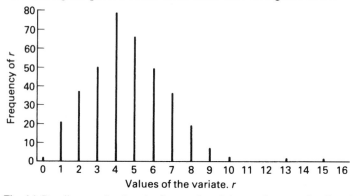

Fig. 6.1 Bar-diagram showing the frequencies of the numbers, r, of radioactive particles emitted per $\frac{1}{2}$ min (Example 6.2.2).

6.2.3 Example
The number of printing errors per page in a random sample of 100 pages from a magazine were:

Number of errors, r:	0	1	2	3	4	5	6	7	Total
Number of pages, f:	8	18	30	25	12	4	1	2	100

To satisfy the Poisson condition (2), we would regard the simple transposition of two letters as one mistake, not two. Exercise 6.5.6 uses these data.

Examples where these conditions seem to be suitable are:

(1)   the number of blood cells of one type in unit volumes of blood;
(2)   the number of radioactive particles emitted from a source in unit time;
(3)   the number of organisms in unit volume of water sampled from a stream or pond;
(4)   the number of vehicles passing per unit time when observed on a road with unrestricted and uninterrupted traffic flows;
(5)   the number of people having a birthday on the same day;
(6)   the number of currants in a currant bun;
(7)   the number of arrivals each five minutes at a doctor's surgery;
(8)   the number of a particular plant species in a sampling quadrat.

In the examples (1) and (3) above, the cell suspensions or the water samples must be kept well stirred to avoid clumping of organisms; if water is taken from the edge of a stream the *negative binomial* (Section 5.7) may explain the data better because it allows for events to clump together rather than arise singly, whereas the Poisson is a perfectly good model for the data from flowing water in mid-stream. In (6), the dough would need to be thoroughly mixed, otherwise the conditions for a Poisson would not apply; in (7), if the patients all arrived by the same public transport rather than individually the Poisson model would also break down.

The mean and variance of the Poisson are equal. In the negative binomial, where some grouping or clumping together occurs, the variance is greater than the mean. The Poisson is often taken as the standard 'random' distribution in ecology, and when the variance is greater than the mean a set of data will be described as 'overdispersed'. In this case, overdispersion is taken as lack of independence: example (8) above may show this if the plants have arrived seed-borne during a short period when the wind was always from the same direction, so that some parts of the sampled area contained large numbers of the plant and others none at all.

6.3 Goodness-of-fit to a Poisson distribution

6.3.1 Example

In a test of electronic components in which the failure rate is thought to be constant, the numbers of failures in a standard test are as follows.

Number of failures, r:	0	1	2	3	4	5	6	Total
Frequency, f	26	30	26	18	9	5	6	120

We may examine whether the data of Example 6.3.1 do appear close to a set of Poisson frequencies. In order to calculate these, a value for μ is required. Usually this is not known and must be estimated from the data. Since μ is the mean, we use $\bar{r}$ in its place.

$$\bar{r} = \frac{(26 \times 0) + (30 \times 1) + (26 \times 2) + (18 \times 3) + (9 \times 4) + (5 \times 5) + (6 \times 6)}{120} = \frac{233}{120} = 1 \cdot 94.$$

The Poisson probabilities therefore come from

$$P(r) = \frac{e^{-1 \cdot 94} (1 \cdot 94)^r}{r!}$$

and the frequencies are these multiplied by 120.

$$P(0) = e^{-1 \cdot 94} \qquad\qquad = 0 \cdot 1435$$

$$P(1) = 1 \cdot 94 e^{-1 \cdot 94} \qquad\qquad = 0 \cdot 2783$$

$$P(2) = \frac{(1 \cdot 94)^2}{2!} e^{-1 \cdot 94} \qquad = 0 \cdot 2700$$

$$P(3) \qquad\qquad = 0 \cdot 1746$$

$$P(4) \qquad\qquad = 0 \cdot 0847$$

$$P(5 \text{ or more}) \qquad = 0 \cdot 0489$$

$$\overline{1 \cdot 0000}$$

These *expected frequencies* can be added to a table of the original data:

r:	0	1	2	3	4	5 or more	Total
f_{OBS}:	26	30	26	18	9	11	120
f_{EXP}:	17·22	33·40	32·40	20·95	10·16	5·87	120·00

We may, on simple inspection, have some doubts about the closeness of fit of 'observed' and 'expected', especially at the two ends (0 and 5 or more). A statistical test using the 'chi-squared' distribution (Chapter 12) can be used to compare these two sets of frequencies; if the fit appears adequate when this test is applied, we may conclude that the Poisson was a satisfactory model and that therefore the conditions for the existence of a Poisson distribution may give a reasonable explanation of the process.

6.4 MINITAB

Once a value for the mean μ has been specified, the probabilities for any r that is requested can be calculated: e.g. if $\mu = 3$ we may ask for $P(2)$ and $P(5)$ and the package will give the answers 0.2240 and 0.1008. When no particular values of r are requested a table is printed which gives all probabilities that are greater than 0.0005. Cumulative probabilities can be found, e.g. $P(r \leq 2) = P(0) + P(1) + P(2)$; these will be set out in a table for values of r beginning at 0 and continuing until the cumulative probability has become 1.0000 to four decimal places.

As for the binomial distribution (Section 5.8) we may find the value of r which has a given cumulative probability; and usually we shall be given two values of r for the same reason as that explained in Section 5.8.

6.5 Exercises on Chapter 6

6.5.1 A radioactive source is emitting, on the average, one particle per minute. If counting continues for several hundred minutes, during which time the particles are emitted randomly, in what proportion of these minutes is it to be expected that

(a) there will be no particle emitted?
(b) there will be exactly one particle emitted?
(c) there will be two or more particles emitted?

6.5.2 The numbers of snails found in each of 100 sampling quadrats in an area were as follows:

Number of snails, r:	0	1	2	3	4	5	8	15
f = frequency of r:	69	18	7	2	1	1	1	1

Find the mean and variance of the number of snails per sampling unit, and use these to gain a rapid idea of whether the data fit the Poisson distribution.

6.5.3 The numbers of plants of a certain species falling in 100 randomly chosen quadrats in an area were counted as follows:

Number of plants	0	1	2	3	4	5	6	7	8	Total
Number of quadrats:	17	20	28	18	8	8	0	0	1	100

Does this suggest that the plants are growing at random in the area?

6.5.4 Random samples, each of unit volume, are taken from a well-mixed cell suspension in which on average one cell of a special type is present per unit volume. Find the probability that a sample will not contain more than one of the special cells.

If 10 such samples are taken, find the probability that exactly five of these will contain not more than one of the special cells.

6.5.5 The average density of a bacterial organism in a liquid is μ per ml. Fifty separate 1 ml units are taken from the liquid, and 10 of these prove to be sterile (i.e. contain no organisms). Find an estimate of μ.

6.5.6 For the data in Example 6.2.3, draw a bar diagram.

Calculate the mean of these data, and calculate also the frequencies of $r = 0, 1, \ldots$ up to '5 or more' to be expected on a Poisson model with the same mean.

Draw these expected frequencies on the same bar diagram (in a different colour, using bars adjacent to those for the observed data) and examine the agreement or otherwise between the two sets of bars.

Finally, calculate the variance of the data. Does it appear to be close enough to the mean for the Poisson model to be satisfactory? (See Clarke and Cooke[3] for a formal test of this.)

6.5.7 The following data are the numbers of a particular type of cell found in unit volumes of blood from animals suffering from a nutritional deficiency.

r:	0	1	2	3	4	5	6	7	8	9	Total
f:	9	15	15	4	2	1	1	1	1	1	50

Calculate the mean and variance, and the expected frequencies on the Poisson model with the same mean as the observed data.

Draw a bar diagram to compare the observed and expected frequencies.

6.5.8 (a) An insect trap catches on average 3·3 insects each quarter-hour on a moonlight night in suitable climatic conditions. Calculate the probability that, in a randomly chosen quarter-hour period:
(i)   four insects will be caught;
(ii)   less than four insects will be caught;
(iii) more than four insects will be caught.

Explain why the model you have used for the counts of insects is a reasonable one. Use it to calculate the probability that no insects are caught in a *five-minute* period.

(b) If the records are now to be taken in five-minute periods calculate:
(i)   the probability that not more than two out of 10 periods will show no insects,
(ii)   the mean number of insect-free periods in samples of 10.

7 The normal distribution as a model for continuous data

7.1 Frequency curves

Suppose that a histogram or a frequency polygon has been drawn as a summary of a set of continuous data, such as heights of people, times to complete a chemical reaction, weights of crops or of plants. It is reasonable to imagine that if we had a very large set of data a *smooth* curve could be drawn to show how frequency $f(x)$ changes as x increases. Continuous (as opposed to discrete) data should vary smoothly, without gaps or jumps in the pattern of variation, and so we can smooth out as well as possible the pattern that may be visible from the histogram or polygon (see Fig. 3.1). After doing this, we look for suitable models to explain the common types of pattern; a statistical model or *distribution* will be expressed in the form of an equation $y = f(x)$, and the mathematical function that is used for $f(x)$ should ideally have any constants or *parameters* in it related as simply as possible to the mean μ and variance σ^2 of the distribution.

7.2 Conditions for the normal distribution

Symmetrical data are very often modelled by the *normal* distribution. If the dot-plot, histogram or stem-and-leaf diagram (Chapter 2) for a set of data seems to satisfy these conditions:

(1) there is a strong tendency to take central values;
(2) variation about the centre is symmetrical;
(3) frequencies of larger deviations decrease rapidly;

then the normal distribution may be used as a model.

The normal curve is shown in Fig. 7.1; it is bell-shaped, and the mean, median and mode (position of maximum frequency) are all equal to μ. The x distance from μ to the point of inflection in the bell shape is equal to σ, the standard deviation. A shorthand for 'the normal distribution with mean μ and variance σ^2' is $N(\mu, \sigma^2)$; note that it is the value of the *variance*, σ^2, not the standard deviation, which is used in this shorthand. The actual equation of the curve (the probability density function) of $N(\mu, \sigma^2)$ is

$$f(x) = \frac{1}{\sigma\sqrt{2\pi}} e^{-\frac{(x-\mu)^2}{2\sigma^2}}, \quad -\infty < x < \infty.$$

This represents a whole family of distributions, all the same basic bell shape. A larger value of μ implies that data are located more to the right on the scale of measurement,

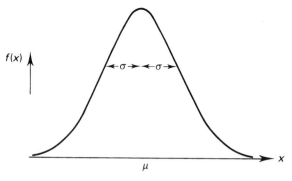

Fig. 7.1 The distribution of frequency, $f(x)$, of values of the variate, x, in the normal distribution with mean μ and variance σ^2.

while larger σ^2 implies a more spread out or scattered set of data (as illustrated in Fig. 3.3). The parameters μ and σ^2 appear directly in the equation to the curve, and so we are able to compare two samples of data from normal distributions by comparing the means and the standard deviations (or the variances). When we are studying a particular problem, or set of data, and using the normal distribution as a model, we must specify values for μ and σ^2 in order to decide which one of the normal family of distributions is required.

Natural data which often follow a normal distribution are those in which the record taken is actually the sum of many parts, e.g. the crop weight on a field plot, where many plants contribute to the measurement, and the height of a human being where the lengths of various parts of the skeleton add together to give the total height. Measurements of this type are sometimes called *additive*. Although the normal distribution is very common, particularly in biological work, we must not immediately assume that almost anything can safely be treated as normal; there are a number of ways of justifying its use, as we shall find in later chapters, but it is none the less foolish to use it without seeing that one of these ways is relevant. If enough data are available to draw a histogram, or dot-plot, or stem-and-leaf diagram and put a fairly smooth curve on top of it we should at least be satisfied that the data seem reasonably near to symmetrical before using the normal as a model for them. The normal distribution was of course discovered in this way by studying large amounts of data. Not all symmetrical sets of data are in fact normal, theoretically speaking, because they may not display the same form of bell shape as the normal with the same values of μ and σ^2; but for the small samples of data that are common in biological work it is not easy to check this, unless one or two observations are very different from the rest, and so we usually take the normal as a suitable model.

Although the equation for the normal curve has no theoretical upper or lower limits for the values of x, the frequency $f(x)$ tails off so quickly either side of μ that virtually all of the distribution is included within the range $x = \mu - 3\sigma$ to $x = \mu + 3\sigma$. There is therefore no objection to using it as a model for physical data that obviously do have limits, such as the heights in human populations or the crop weights from field plots.

The two mathematical constants $e = 2.718\ 28\ \ldots$ (see Section 6.2) and $\pi = 3.141\ 59$ $\ldots$ both appear in the equation; it is e which helps to determine the shape of the curve, and the expression $\frac{1}{\sigma\sqrt{2\pi}}$ serves to make the area under the curve, inside the bell shape, equal to 1. As we shall see (Section 7.3), we need to study the proportions of this total area which are associated with specified values of x.

Every field of application of statistics contains examples of data which follow the normal distribution to at least a good degree of approximation. In industry, the lengths of mass-produced components when they emerge from a production line will often vary about the target value according to a normal distribution; and some of the standard I.Q. tests used in educational assessment produce scores which follow a normal distribution, as do the tests used by personnel officers in assessing the skills needed for certain types of work. Thus the methods we shall go on to study in this text have wider application than the biological and agricultural sciences, although this is where many of them were first discovered and applied.

7.3 The normal distribution in statistics

The normal distribution also often arises as an approximation to other distributions (see Section 7.5) and therefore statistical methods developed for the normal distribution can often be used more widely without having to develop new theory and methods for every new distribution. This is particularly true where large samples of data are available, and their means are being studied. It is a very important theoretical distribution indeed; Gauss studied it in physics and astronomy, and it is sometimes called *Gaussian*. The name *normal* seems to have been coined by the British biometrician K. Pearson, and has been common since the beginning of this century.

A very important normal distribution is the *standard* (unit) normal $N(0,1)$, with mean 0 and variance 1, usually denoted by z. It is extensively tabulated, giving the probabilities up to a particular value of z, i.e. the cumulative probability for z. As in a histogram, it is the *area* standing on a particular interval under a continuous curve which represents probability in that interval.

Tables usually have entries for $z \geqslant 0$ and rely on symmetry for $z < 0$ (see Table VII). We shall not need to be concerned with the equation of the normal curve, only with the information given in Table VII. We can use it to find the probability of obtaining values of z that are less than some specified value Z. Then we may use the argument that *either* z is less than Z, *or* z is greater than or equal to Z, because together these two possible events account for all of the distribution; therefore $P(z < Z) + P(z \geqslant Z) = 1$. Examples 7.3.1 to 7.3.6 illustrate various types of calculation that may be needed when the normal distribution is used as a model for data.

7.3.1 Example
In $N(0, 1)$, $P(z < 1 \cdot 5) = 0 \cdot 9332$, from Table VII. Therefore $P(z \geqslant 1 \cdot 5) = 1 - 0 \cdot 9332 = 0 \cdot 0668$.

A diagram is always useful when evaluating probabilities in a normal distribution, and Fig. 7.2(a) illustrates this example.

7.3.2 Example
Because the distribution is symmetrical, $P(z < -Z)$ is the same as $P(z \geqslant +Z)$; hence, as shown in Fig. 7.2(b), $P(z < -1 \cdot 5)$ is equal to $P(z \geqslant +1 \cdot 5) = 0 \cdot 0668$.

7.3.3 Example
The probability between two values of z is found by looking up the two table entries

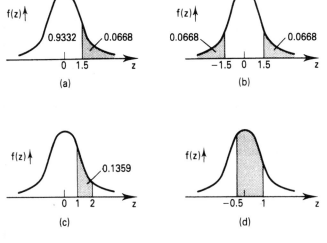

Fig. 7.2 Diagrams illustrating (a) Example 7.3.1, (b) Example 7.3.2, (c) Example 7.3.3, (d) Example 7.3.5 in the standard normal distribution $N(0, 1)$.

and subtracting. For example, $P(1 < z < 2) = P(z < 2) - P(z < 1)$, which from Table VII is equal to $0.9772 - 0.8413 = 0.1359$. Figure 7.2(c) illustrates this.

7.3.4 Example
The symmetry method of Example 7.3.2 can also be used to find $P(z > Z)$ when Z takes a negative value; for example $P(z > -0.75)$ is, by symmetry, equal to $P(z < +0.75)$, which from Table VII is 0.7734.

7.3.5 Example
In a similar way we can tackle the more complicated problem of finding $P(-0.5 < z < 1)$. Again a diagram should be drawn, and Fig. 7.2(d) shows this.

$$P(-0.5 < z < 1) = P(z < 1) - P(z < -0.5).$$

First calculate $P(z < -0.5)$, which is equal to $P(z > +0.5)$ by symmetry. This is $1 - P(z < +0.5) = 1 - 0.6915$ from Table VII, and finally this equals 0.3085. Also from Table VII, $P(z < 1) = 0.8413$, and therefore $P(-0.5 < z < 1) = 0.8413 - 0.3085 = 0.5328$.

7.3.6 Example
For later use (Chapters 8 onwards) we shall want to know the z-values between which probabilities such as 0.90, 0.95, 0.99, 0.999 lie. Now from Table VII, $P(z < 1.96) = 0.9750$, and so $P(z \geqslant 1.96) = 0.0250$; by symmetry also $P(z < -1.96) = 0.0250$ and therefore $P(-1.96 < z < +1.96) = 0.95$. The central 95% of $N(0, 1)$ lies in the range $(-1.96, +1.96)$. Figure 7.3(b) shows this result, and Fig. 7.3(a) shows other useful values for significance testing and the calculation of confidence intervals, which we shall consider in Chapters 8 and 9.

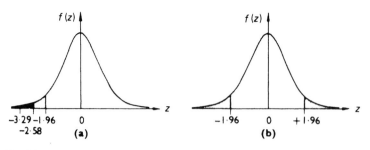

Fig. 7.3   (a) Values $Z = -3.29$, -2.58, -1.96 in the $N(0, 1)$ distribution, whose cumulative frequencies are respectively $\frac{1}{20}\%$, $\frac{1}{2}\%$, $2\frac{1}{2}\%$. (b) Values $Z = \pm 1.96$ which enclose the central 95% of the $N(0, 1)$ distribution.

7.4 Standardisation

Clearly it would be impossible to prepare tables of every possible normal distribution— there is an infinite number of them! For example, we may wish to consider two alternative models for some data which can be expressed in normal distributions with the same variance: this variance is 5, but the two means are different. The first is 4, so that this model is the $N(4, 5)$ distribution, and the second is 7, leading to $N(7, 5)$. When we draw graphs of $f(x)$ against x, the second will be further to the right on the scale of measurement because it has a larger mean (7) than the other one (4). This type of situation was illustrated by the two curves (α) and (β) of Fig. 3.3. Similarly curves (β) and (γ) of Fig. 3.3 show what would happen if the two sets of data were to be modelled by $N(7, 5)$ and $N(7, 10)$: the second of these, with variance 10 would be a much wider curve than the first in which the variance was 5, though both have the same mean (7).

We can *standardise* any $N(\mu, \sigma^2)$ into $N(0, 1)$, which is called the *standard normal distribution*, as follows. If X is any member of the normal family, say $N(\mu, \sigma^2)$, subtract the mean μ from every observation and divide the result by σ, to give $Z = (X - \mu)/\sigma$ which will have mean 0 and variance 1. Z can now be looked up in tables because it is still normally distributed; $N(0, 1)$ in fact. So on the z scale we are actually considering deviations from μ, which we have made the new origin of our measurements; and the standard deviation σ has become the new unit of measurement. Care must be taken to include the minus sign when we have a value of x that is less than μ.

7.4.1 Example

If X is taken from $N(0.5, 16)$, then $Z = \frac{X-0.5}{4}$ is $N(0, 1)$. The value $x = 1.928$ corresponds to

$$Z = \frac{1.928 - 0.5}{4} = 0.3570.$$

Therefore the probability that X is less than 1.928 in $N(0.5, 16)$ is equal to the probability that Z is less than 0.3570 in $N(0, 1)$, i.e. 0.6395. The probability that X is > 1.928 is then $1 - 0.6395 = 0.3605$.

7.4.2 Example

Find $P(3 < X < 4)$ when X is distributed $N(1, 4)$. Each value of X has to be standardised by calculating $z = \frac{X - \mu}{\sigma}$ which is $z = \frac{(X-1)}{2}$. For $X = 3$ this gives $z = 1$, and for

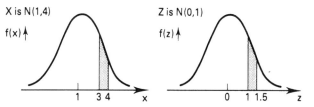

X is N(1,4)   f(x)↑   1 3 4   x

Z is N(0,1)   f(z)↑   0 1 1.5   z

Fig. 7.4   Equivalent values of X which is $N(1,4)$ and Z which is $N(0,1)$

$X=4$ we obtain $z=1.5$; thus the required probability is $P(1<z<1.5)$, which is $P(z<1.5)-P(z<1)=0.9332-0.8413=0.0919$. Two diagrams, the first showing the values in the original distribution x and the second showing the corresponding standardised values z, will be of great help in making sure that the right probability is calculated; Fig. 7.4 provides these for Example 7.4.2.

7.4.3 Example

Extending Example 7.3.6, suppose that x is $N(\mu, \sigma^2)$. Then the two values $x=\mu+1.96\,\sigma$ and $x=\mu-1.96\,\sigma$ standardise into

$$Z = \frac{(\mu + 1.96\sigma) - \mu}{\sigma} \quad \text{and } Z = \frac{(\mu - 1.96\sigma) - \mu}{\sigma}$$

which are simply $z=1.96$ and $z=-1.96$. Therefore in $N(\mu, \sigma^2)$ the central 95% lies in $(\mu-1.96\,\sigma,\ \mu+1.96\,\sigma)$.

7.5 Transformation to normality

Examples also exist of variates which are not themselves normal, but can be made so by a suitable *transformation* of their scale of measurement. When an insecticide is applied experimentally to insects, the concentration x of insecticide needed to kill any particular insect—what is called the tolerance of that insect—is not constant for all insects in the experiment; in fact it often happens that when the frequency of kill is plotted against concentration, it gives a skew curve like Fig. 7.5(a). By plotting log x, instead of x, on the horizontal axis, the curve can be normalised (Fig. 7.5(b)), and we say that the original data on concentration, x, and kill were *log-normally* distributed, since log x is normal. (Note that $x=1$ in Fig. 7.5(a) corresponds to log $x=0$ in (b).) The log-normal distribution is also found in metallurgy, where specimens have been cast from an alloy and are subjected to a test for a time x before showing cracks or other symptoms of metal fatigue.

Although it may at first sight look like a mathematical trick, there is nothing 'unnatural' about taking measurements on a logarithmic scale. Many statistical (and other) methods of comparing sets of data look at *differences* between means etc.; sometimes it is much more appropriate to consider *ratios*. When an organism is growing at a rate which is proportional to its present size, its weight is changing on a *multiplicative* scale, not an additive one—'exponential' growth. In later chapters, we shall examine the 'location' of a distribution, or compare two distributions, using the mean and differences between means. The logarithm of a ratio, log (v/w), is equal to

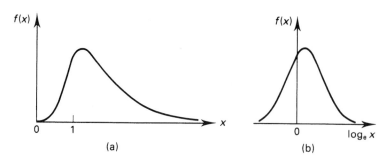

Fig. 7.5   The log-normal distribution: (a) shows $f(x)$ plotted against x and (b) shows $f(x)$ plotted against $\log_e x$. (The distribution in (b) is normal.)

the difference of the logarithms, $\log v - \log w$. So the use of a logarithmic scale enables ratios to be examined using the same theory as that developed for differences. Because so much theory exists for the normal distribution, it is worth trying the lognormal as a possible model when dealing with skew data; but of course not every skew shape can be modelled in this simple way and if a lognormal does not fit well then nonparametric methods (Chapter 10) may be best.

7.6 Approximate normality

The normal distribution can also be used to approximate to several other standard distributions, given suitable conditions on their parameters: we say that these distributions *converge to normality*. It is even possible to approximate to discrete distributions in this way, in spite of the normal being itself continuous. The *convergence* is in the sense that, when the necessary conditions are satisfied, a frequency graph for the actual distribution will be very similar indeed in shape to the normal which is used as its approximation. Hence statistical summary measures and significance tests (Chapters 8 and 9) worked out for the approximating normal distribution may be used, instead of carrying out the calculations for the actual distribution itself. This is often easier to do, and also has the advantage that separate methods and theory need not be developed for each distribution. Examples of the required conditions are that a *binomial* distribution can be approximated, for large n, by $N(np, np(1-p))$, the normal with the same mean and variance as the binomial; and a *Poisson*, for large μ, is very close to $N(\mu, \mu)$. For the binomial, the approximation is satisfactory for $n > 100$, and also for smaller values of n (say > 30) if p is near to $\frac{1}{2}$. (For $p = \frac{1}{2}$, the binomial is in any case a symmetrical distribution, but this does not automatically imply normality. The rules for approximation tell us how large n needs to be before a bar-diagram of the binomial would be covered very closely by the curve of its approximating normal.) For the Poisson distribution, the approximation is a good one when μ is 5 or greater. In both these cases, binomial and Poisson, the approximating normal distribution has the *same* mean and variance as the distribution being approximated to; this is a perfectly general rule for all situations where a normal approximation exists.

   It should be realised that when we summarise a set of data by quoting its mean and standard deviation, we are in a sense assuming it to be, more or less closely, normally distributed: for if we know that the data follow a binomial or a Poisson distribution, other parameters are appropriate (n and p, or μ respectively), and if we know the distribution to be rather skew, or of no particular shape at all, there is often a case for using medians and the measures of dispersion related to them. It *is* very common to summarise data by mean and standard deviation, but a moment's thought should be given to the appropriateness of this.

7.6.1 Example

Because the binomial distribution is discrete, it is specified by giving a formula for $P(r)$, where r takes the values 0, 1, 2, . . . , n (Section 5.5). The normal approximation cannot do this because the normal distribution is continuous, not discrete. *We must approximate $P(r)$ in the binomial by the probability between $r-\frac{1}{2}$ and $r+\frac{1}{2}$ in* $N(np, np(1-p))$.

(a) If a random sample of $n=20$ items is taken from a binomial distribution in which the probability of one special type of member is $p=\frac{3}{8}$, the binomial formula gives (for example):

$$P(7) = \frac{20!}{7!13!}\left(\frac{3}{8}\right)^7\left(\frac{5}{8}\right)^{13} = 0\cdot1795,$$

after a rather tedious calculation.

The normal approximation is $N(20 \times \frac{3}{8}, 20 \times \frac{3}{8} \times \frac{5}{8})$, i.e. $N(7.5, 4\cdot6875)$. The value for $P(7)$ is approximated by the probability between 6.5 and 7.5 in this distribution. If X is $N(7\cdot5, 4\cdot6875)$ then $Z=\frac{(X-7\cdot5)}{\sqrt{4\cdot6875}}$ is $N(0, 1)$, the standard normal distribution. When $X=6\cdot5$, $Z=\frac{(6\cdot5-7\cdot5)}{\sqrt{4\cdot6875}}=\frac{-1}{2\cdot165}=-0\cdot4619$; and when $X=7\cdot5$, $Z=0$. The required probability is $P(-0\cdot4619<z<0)=P(z<0)-P(z<-0\cdot4619)$. By the method of Example 7.3.2, $P(z<-0\cdot4619)=P(z\geqslant+0\cdot4619)=1-P(z<0\cdot4619)=1-0\cdot6779=0\cdot3221$. This gives the required probability as $0\cdot5000-0\cdot3221=0\cdot1779$ which is a very good approximation.

(b) In the same binomial, to find $P(10\leqslant r\leqslant15)$ would be even more tedious a calculation, requiring $P(10)+P(11)+P(12)+P(13)+P(14)+P(15)$. By the normal approximation it reduces to finding $P(9\cdot5\leqslant x\leqslant15\cdot5)$ in the distribution $N(7\cdot5, 4\cdot6875)$. When $x=9\cdot5$, $z=\frac{9\cdot5-7\cdot5}{\sqrt{4\cdot6875}}=0\cdot9238$ and when $x=15\cdot5$ then $z=\frac{(15\cdot5-7\cdot5)}{\sqrt{4\cdot6875}}=3\cdot6950$. The required probability is $P(0\cdot9238\leqslant z\leqslant3\cdot6950)=P(z<3\cdot6950)-P(z<0\cdot9238)=0\cdot9999-0\cdot8222=0\cdot1777$.

7.6.2 Example

The normal approximation to a Poisson distribution is especially useful when we require the upper tail probability $P(r\geqslant R)$, the calculation of which is complicated by the fact that r has no upper limit. Suppose we are studying a random process in which events occur at the rate of eight per hour, and we wish to find the probability that in one hour there will be 12 or more events. One way of doing so would of course be to calculate all twelve probabilities $P(0)$ up to $P(11)$, add these together and take the result away from 1. The normal approximation is a great improvement on this.

The Poisson with $\mu=8$ is approximated by the normal distribution with the same mean and variance, $N(8, 8)$. Therefore if r is the number of events per hour then $z=\frac{(r-8)}{\sqrt{8}}$ is $N(0, 1)$.

Again we need the *continuity correction* of moving half a unit from r in the calculation: $P(r \geqslant 12)$ is found by taking the normal approximation from 11·5 upwards. When $r = 11·5$, $z = \frac{(11·5-8·0)}{\sqrt{8·0}} = \frac{3.5}{2·8284} = 1·2374$. $P(z \geqslant 1·2374) = 0·1080$ which is the required answer.

If now we require the probability that r is more than 12, *but not including* 12, that is $P(r > 12)$ rather than $P(r \geqslant 12)$, the upper tail of the normal approximation must begin at 12·5, *not* 11·5; we do not want the probability for $r = 12$ to be included in the answer. This time we need $z = \frac{(12·5-8·0)}{\sqrt{8·0}} = \frac{4.5}{2·8284} = 1·5910$, corresponding to $r = 12·5$. The probability $P(z \geqslant 1·5910)$ is 0·0558.

From these two calculations, we see that $P(r = 12) = 0·1080 - 0·0558 = 0·0522$. If we had used the Poisson formula directly to calculate $P(r = 12)$ it would have been $e^{-8}8^{12}/12!$ which gives 0·0481.

7.6.3 Example
A radioactive source is emitting particles at the rate of two per minute, but the time unit for recording is five minutes. What is the probability of recording less than eight particles in this time?

The mean rate of emission per 'unit' time (five minutes) is 10 particles. The number r emitted may be approximated by $N(10, 10)$, and allowing for the continuity correction we require $P(r < 7\frac{1}{2})$. The z-value corresponding to $7\frac{1}{2}$ is $\frac{(7\frac{1}{2}-10)}{\sqrt{10}} = -0·791$.

$$P(z < -0·791) = P(z > 0·791) = 1 - P(z < 0·791) = 1 - 0·7855 = 0·2145.$$

7.6.4 Example. Proportions
Often in a population containing two types it is not the distribution of the actual number r of special type which is interesting, but the *proportion* of special type found in a sample. We shall anticipate a later notation (Chapter 9) and call this sample proportion $\hat{p}$ ('p-hat') because it is an estimate of p in the population from which the sample was drawn. So $\hat{p} = r/n$, in the notation used for the binomial distribution.

Now we know that r is binomial, with mean np and variance $np(1-p)$, so that under the usual conditions on n and p it can be approximated by $N(np, np(1-p))$. When we divide r by n, to obtain $\hat{p}$, r must have its mean divided by n; and because the variance is measured in units which are the squares of the actual observations it must be divided by n^2. The distribution of $\hat{p}$ can still be approximated by a normal distribution, since in dividing by a constant number n we do not alter the shape; in this approximation the mean will be $np/n = p$, and the variance will be $np(1-p)/n^2 = p(1-p)/n$. This gives the very important result that, for $n > 30$ if p is near $\frac{1}{2}$, or in general for $n > 100$, the distribution of a sample proportion $\hat{p}$ is approximately $N(p, p(1-p)/n)$, where p is the true proportion in the whole population.

(a) Two hundred seedlings are planted out from a nursery to a permanent site. The experimenter knows that usually 80% will survive. What is the probability that no more than 25 will die?

The conditions for a normal approximation are satisfied, and the approximating distribution has mean $np = 200 \times 0·8 = 160$ and variance $200 \times 0·8 \times 0·2 = 32$. The number r which survive is to be at least 175. Find the probability of exceeding $174\frac{1}{2}$ in $N(160, 32)$.

$$z = \frac{(174\frac{1}{2}-160)}{\sqrt{32}} = 2·56 \text{ which has probability 0.9948 up to it so the required probability is}$$
$1 - 0·9948 = 0·0052$.

(b) If the nurseryman who raises the seedlings resows his empty ground with seeds that have a germination rate of 60%, and places them in blocks of 500 seeds, what is the minimum proportion that he can expect, with probability 0·95, to germinate in a block?

The approximating normal distribution for $\hat{p}$ has mean 0·6 and variance $\frac{(0.6 \times 0.4)}{500} = 0.000\,48$. The z-value which excludes the lowest 5% of values is -1.645. Therefore $z = \frac{(\hat{p} - 0.6)}{\sqrt{0.000\,48}} = -1.645$, giving $\hat{p} = 0.06 - 1.645 \times 0.0219 = 0.564$ as the minimum expected proportion germinating. (This is 282 out of 500.)

If he used blocks of 50 seeds, his results would be more variable, since now the variance of $\hat{p}$ is $\frac{(0.6 \times 0.4)}{50} = 0.0048$, and $\sqrt{0.0048} = 0.0693$, giving the minimum likely value of $\hat{p}$ as $0.6 - 1.645 \times 0.0693 = 0.486$. (This is 24 out of 50.)

Note. 1. To make normal approximations to binomial distributions as good as possible for extreme values of p, less than 0·1 or greater than 0·9, it is wise to take n at least 250. It is also desirable in this case to observe an additional rule, namely that np or $n(1-p)$, as appropriate, should be greater than 5. This rule helps to ensure that we do not try to approximate a binomial distribution which is skew by a normal distribution which is symmetrical.
2. In a continuous distribution, $P(x < X)$ and $P(x \leqslant X)$ are the same: probabilities can only be defined in intervals, not at points. This is the opposite of the case for a discrete distribution; as we have seen, it matters very much whether or not a particular value of r is to be included in the calculation.

7.7 Sampling distributions and the central limit theorem

The **Central Limit Theorem** is the most common justification for using the normal distribution, and to appreciate it we need the idea of a *sampling distribution*. Let us suppose that we take a sample of n observations, at random, from *any* frequency distribution, not necessarily normal: we ask of it only that the frequency $f(x)$ shall tail off sufficiently rapidly as x increases for it to possess a finite mean and variance (in practice we could assume this for biological data in general). This sample will have a mean, $\bar{x}$. Suppose we now take a second sample, also of n observations, from the same original *parent* distribution; this too will have a mean, whose numerical value will be slightly different from the first $\bar{x}$ because the sample, although from the same parent population, consists of different members. As we go on taking samples, so we go on getting different numerical values for their means, and this set of numerical values represents the *sampling distribution of the mean*. At first sight we might suppose that the shape or pattern of this sampling distribution would depend very much on the shape of the parent distribution; but this is not so. It is a very remarkable mathematical result that the sampling distribution of the mean of samples from *any* parent distribution approaches a normal distribution very closely as sample size n increases. Further, we can say what the mean and variance of this sampling distribution of means will be: if μ, σ^2 denote the parent mean and variance, then the sample mean based on n observations has the distribution of $N(\mu, \sigma^2/n)$, this being an approximation satisfied more and more closely as n increases.

When the parent distribution is itself *normal*, the sampling distribution of means is *exactly* normal for all sample sizes n. This is the basis of the rule saying that the standard deviation of the mean of a sample is $\sigma/\sqrt{n}$, and shows that the precision of determination of a mean improves directly as the square root of the number of observations upon which it was based: its distribution becomes narrower as n increases, as shown in Fig. 7.6.

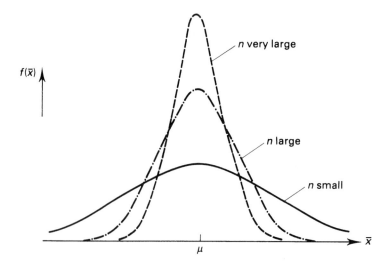

Fig. 7.6   Distribution of sample mean $\bar{x}$ about its true value μ for varying sample sizes n.

In a Poisson distribution with mean μ greater than 5, the normal distribution $N(\mu, \mu)$ approximates $P(r)$ as explained in Section 7.6. Therefore if the mean, $\bar{r}$, of n observations from a Poisson distribution is being studied we can use $N(\mu, \frac{\mu}{n})$ to approximate the distribution of $\bar{r}$.

7.7.1 Example
The mean increase in diameter of a fungal spot incubated under standard conditions for a fixed period of time is 14·5 mm. The distribution of diameters is assumed to be normal, and its variance is 0·8 mm². A random sample of eight spots grown under these conditions is selected. The probability that the mean increase in diameter of these eight spots is less than 14·0 will be the probability of finding a value less than 14·0 in a normal distribution with mean 14·5 and variance $\frac{0·8}{8} = 0·1$. This is the probability that, after standardization, z is less than $\frac{(14·0 - 14·5)}{\sqrt{0·1}} = -\frac{0·5}{0·316} = -1·581$, which is equal to

$$P(z > +1·581) = 1 - P(z < 1·581) = 1 - 0·9430 = 0·057.$$

7.7.2 Example
The arrival times of workers at a certain company follow a normal distribution, with mean at 8.50 a.m. and standard deviation five minutes. The probability of a 'late' arrival, if 9 a.m. is the official starting time, is therefore the probability that X is greater than 10 in $N(0, 5^2)$, i.e.

$$P\left(z > \frac{10}{5}\right) = P(z > 2) = 1 - P(z < 2) = 1 - 0·9772 = 0·0228.$$

For a whole week (of five days), supposing the reasons for late arrival are purely random, the *average* arrival time for a worker will be normally distributed about 8.50 a.m. with variance $\frac{5^2}{5}$, i.e. standard deviation $\sqrt{5}$. So the probability of late arrival on average is the probability that X is greater than 10 in $N(0, 5)$, i.e. $P(z>4\cdot47)$ which is effectively zero. In this average, late days are balanced by early ones.

7.8 Combining normal distributions

When two or more sets of data can be modelled by normal distributions, there are some very useful and important results that allow sums or differences of measurements to be studied. Suppose that X is $N(\mu_1, \sigma_1^2)$ and Y is $N(\mu_2, \sigma_2^2)$; and X is independent of Y.
(a) The sum $X + Y$ is also normally distributed. Its mean is $(\mu_1 + \mu_2)$ and its variance is $\sigma_1^2 + \sigma_2^2$.

7.8.1 Example
The time for the first phase of a reaction to be completed is distributed as $N(12, 3)$ minutes, and that for the second phase is $N(8, 1\cdot4)$ minutes. (Remember that 3 and $1\cdot4$ are the *variances*, not standard deviations.) The total time to complete the reaction will therefore be distributed as $N(12+8, 3+1\cdot4)$, that is $N(20, 4\cdot4)$.

The probability of requiring more than 22 minutes for the whole reaction can now be found. The standardised z corresponding to 22 in $N(20, 4\cdot4)$ is $z = \frac{(22-20)}{\sqrt{4.4}} = 0\cdot953$.

$$P(z > 0\cdot953) = 1 - P(z < 0\cdot953) = 1 - 0\cdot8297 = 0\cdot1703.$$

(b) The difference $X - Y$ is, again, normally distributed. It has mean $\mu_1 - \mu_2$ and variance $\sigma_1^2 + \sigma_2^2$. *Variances are always added, never subtracted.*

7.8.2 Example
A food product is packed in cartons of nominal weight 1000 g. The distribution of weights is normal, and the variance is 1 g. In order to avoid too many under-weight cartons being produced, the mean is set at 1002 g.

The probability that two cartons differ in weight by more than 1 g is found from the distribution of $X - Y$, which is $N(0, 2)$ since both X and Y are taken from $N(1000, 1)$ and are independent samples from that distribution. Note that in this problem *both* ends of the distribution are to be included; X may be greater or less than Y.

We require $P(X - Y) > +1$ and $P(X - Y) < -1$. These will be equal, by symmetry, since the mean of $(X - Y)$ is 0. The standardised value corresponding to $+1$ is $z = \frac{(1-0)}{\sqrt{2}} = 0\cdot7071$, and from Table VII $P(z < 0\cdot7071) = 0\cdot7602$. We want $P(z > 0\cdot7071)$, which is $1 - P(z < 0\cdot7071) = 1 - 0\cdot7602 = 0\cdot2398$. By symmetry, $P(z < -0\cdot7071)$ is also $0\cdot2398$, and the combined probability of the two tails is $2 \times 0\cdot2398 = 0\cdot4796$.

If the probability of differing by more than 1 g is $0\cdot4796$, then the probability of *not* doing so is $1 - 0\cdot4796 = 0\cdot5204$.

As usual, a diagram helps to ensure that the correct areas are found.

(c) If X, Y are multiplied by constants (numbers) a, b and added, the resulting distribution is still normal: $W = aX + bY$ is normal with mean $a\mu_1 + b\mu_2$ and variance $(a^2\sigma_1^2 + b^2\sigma_2^2)$.

7.8.3 Example
A course is assessed by combining scores on a test paper X and laboratory work Y. It is thought that X and Y can be modelled by normal distributions. The overall assessment

mark is taken to be $W = 3X + 2Y$. If it is reasonable to regard X and Y as independent of one another, the distribution of W will be normal with mean $3\mu_1 + 2\mu_2$ and variance $9\sigma_1^2 + 4\sigma_2^2$. Thus if the means of X and Y are 35 and 28 respectively, and the variances in these distributions are 12 and 6, the distribution of W will be $N(161, 132)$. All the types of calculation already illustrated can now be performed on W.

(d) The total of n observations forming a random sample from $N(\mu, \sigma^2)$ also follows a normal distribution, whose mean is $n\mu$ and variance $n\sigma^2$: this is really only an extension of (a).

7.8.4 Example

The weights of animals from a certain population are normally distributed with mean 16 kg and variance 10 kg^2. The total weight of six animals selected at random from this population will then be distributed $N(6 \times 16, 6 \times 10)$, i.e. $N(96, 60)$.

Suppose these animals are to be transported in a vehicle. The probability that their total weight will exceed 120 kg is the probability that the value 120 is exceeded in $N(96, 60)$, and this is standardised in the usual way to give $z = \frac{(120-96)}{\sqrt{60}} = \frac{24}{7\cdot746} = 3\cdot098$. From Table VII,

$$P(z < 3\cdot098) = 0\cdot9990 \text{ and so } P(z > 3\cdot098) = 1 - 0\cdot9990 = 0\cdot0010.$$

Note that this is different from the distribution of six times the weight of one animal: by rule (c) that would follow $N(6 \times 16, 36 \times 10)$, or $N(96, 360)$. When adding the weights of six animals together, large and small weights tend to cancel one another out and so the variance is smaller than that for six times the weight of a single animal.

7.9 MINITAB computing

7.9.1 Symmetry

As indicated in Section 7.5, the appropriate units in which a measurement x may be studied could be logarithmic, using $\log x$ instead of x; and in Chapter 18 we shall see that the square root $\sqrt{x}$ is also useful. Various other possibilities, such as $-\frac{1}{x}$, have been suggested. There is a facility in MINITAB to examine the 'histogram' of a set of data (which is in fact a stem-and-leaf diagram (Section 2.6)) in the original units x and then in any or all of the units $\log x$, $\sqrt{x}$, $-\frac{1}{x}$. The 'histograms' may be compared, and perhaps also the closeness (or otherwise) of means and medians noted, as a guide to which units give the most symmetrical picture. The data could then be analysed in these units as though they were (approximately) normal.

7.9.2 The normal distribution

A normal distribution can be fitted to a set of data. Probabilities must be computed for a suitable set of intervals. Given a set of data ranging in value from A to B, we can split (A, B) into at least 10 intervals, all the same width for simplicity, and specify these within the program for calculating probability density functions. The instruction {NORMAL, μ, σ} where μ and σ are the mean and the standard deviation of the data, will give the probabilities to be plotted at the centre of each interval; joining these plots produces the curve $N(\mu, \sigma^2)$.

7.9.3 Normal approximation to binomial (or Poisson) distribution

The probabilities for a binomial (or Poisson) distribution and its approximating normal distribution can be plotted together on the same diagram, using a command

MPLOT for handling two sets of data. By having both on the same diagram it is easy to decide whether the approximation is sufficiently close.

7.9.4 Normal probabilities
Cumulative probabilities (like those in Table VII for $N(0, 1)$) can be found for any $N(\mu, \sigma^2)$; the values of μ and σ must be specified. The inverse of this process is also possible, giving the value of X below which some stated proportion of the distribution will lie. These facilities provide the necessary calculations for many of the methods of later chapters, avoiding the need to look up printed tables.

7.10 Exercises on Chapter 7

7.10.1 Find the unit (standard) normal deviate corresponding to:
  (a) the value $5\cdot00$ in $N(3\cdot95, 2\cdot25)$;
  (b) the value $0\cdot29$ in $N(0\cdot50, 0\cdot64)$;
  (c) the value $-0\cdot47$ in $N(1\cdot38, 1\cdot21)$;
  (d) the value $-6\cdot89$ in $N(-6\cdot50, 0\cdot04)$.

7.10.2 Draw rough graphs to show how the following pairs of normal distributions differ from each other:
  (a) $N(3, 1)$ and $N(3, 4)$;
  (b) $N(0, 2)$ and $N(6, 2)$;
  (c) $N(4, 1)$ and $N(-4, 1)$;
  (d) $N(0, 1)$ and $N(0\cdot5, 0\cdot1)$.

7.10.3 Sixty-four observations are selected at random from $N(10, 25)$, and their mean $\bar{x}$ is calculated. What distribution will $\bar{x}$ follow?
  What is the probability that $\bar{x}$ will be less than 9?

7.10.4 The nitrogen content of the leaves of a population of strawberry plants at a particular stage of growth varies according to a normal distribution with mean $2\cdot43$ ppm and standard deviation $0\cdot025$ ppm.
  Find the probability that one plant selected at random from this population will have nitrogen content between $2\cdot40$ and $2\cdot41$ ppm.
  Find also the probability that the average nitrogen content in a random sample of 10 plants from this population will be between $2\cdot43$ and $2\cdot44$ ppm.

7.10.5 A measurement x cm is made in an experiment, and is repeated n times. The standard deviation of x is $0\cdot5$ cm. What should n be if the standard deviation of the mean measurement $\bar{x}$ is to be $0\cdot1$?

7.10.6 A set of weights, measured in lb, is distributed as $N(10, 14\cdot65)$. What will be the distribution of the same weights when they are expressed in kg? [1 lb $= 0\cdot4536$ kg].

7.10.7 A manufacturer is making a large supply of a particular item for use in a laboratory. His production machinery is giving a normal distribution of the masses of the items. He finds that the total mass of 100 of the items is 45 551 g, and the estimated variance of mass is $33\cdot33$ g^2. The target mass which the manufacturer is aiming to make is 454 g. Estimate the probability that an item has mass less than 454 g.

7.10.8 A farmer is assessing the cost of using a new piece of machinery to harvest a crop. The main items of cost are fuel and wages. The salesman produces figures to show that on average it should take 25 litres of fuel and four man-hours of labour to complete the harvest in a field the size he has. If it could be assumed that the distributions of fuel consumption and labour time were both normal, and the costs of one litre of fuel and one man-hour of work are both known, what further information would be needed to specify the distribution of total cost completely? Given this information, what would be the distribution of the total cost of completing the harvest?

7.10.9 Fertiliser is packed in (nominally) 5 kg bags. The packer has set his machinery to give bags of average weight 5·1 kg, and he knows that the distribution of bag weights will be normal with standard deviation 0·05 kg. A farmer buys 10 bags at a time. What is the distribution of mean weight of bags per purchase?

7.10.10 The standard method of growing a certain flower crop produces flowers whose diameters have a normal distribution with mean 13 cm and standard deviation 1·2 cm. Flowers with diameter above 15·5 cm are called Grade 1 blooms and command a higher price. Experiments have shown that an increase in the spacing of the crop increases the mean bloom diameter by 1·2 cm, though the standard deviation is unchanged. Calculate the change in the proportion of Grade 1 blooms if the increased spacing distance is adopted.

7.10.11 Which of the following binomial distributions could be approximated closely by a normal?

(a) the binomial $n = 8$, $p = \frac{1}{6}$;
(b) the binomial $n = 80$, $p = \frac{3}{5}$;
(c) the binomial $n = 50$, $p = \frac{1}{10}$;
(d) the binomial $n = 300$, $p = \frac{9}{10}$.

State the means and variances of the appropriate normal distributions.

7.10.12 Two hundred observations are selected at random from a distribution (of unknown type) whose mean is 5 and variance 8. What distribution will the mean, $\bar{x}$, of 200 observations follow? What could you say about $\bar{x}$ if only 20 observations were available instead of 200?

7.10.13 An audit of an area agricultural office revealed 120 out of 500 randomly chosen records relating to smallholdings in the region to be inaccurate in some minor detail. Using this as an indication of the overall proportion of incorrect records in the office, what is the probability that a research worker using another 100 of these records in a survey will find that less than 15 of them contain any error?

7.10.14 The number of items of computer equipment needing maintenance service in a large research institution over a six-month period was 72 out of a total of 396 items in use. What distribution will model these data? Assuming that performance does not change during the next six months, what is the probability that between one-quarter and one-third of the items will need service?

7.10.15 A machine dispenses an amount x of a liquid each time it is activated. Over a long period of steady operation, 4% of sampled output has values of x less than or equal to 99 ml, and $22\frac{1}{2}\%$ has values of x greater than or equal to 104 ml. Assuming that x follows a normal distribution $N(\mu, \sigma^2)$, estimate μ and σ.

8 Significance tests for means: normal and Student's t

8.1 Significance testing

Often a set of data is collected, or an experiment carried out, not simply with a view to summarising the results and estimating suitable parameters but rather in order to *test an idea*. This idea or *hypothesis* may arise by purely theoretical reasoning, or it may have been suggested by the results of earlier experiments. In Chapter 6 we argued that if a set of data, such as the counts of numbers of insects or of radioactive particles, had been generated by a random process, the data should conform to a Poisson distribution. Thus if a set of such data does follow the Poisson distribution, it could have been generated in this way; but if it does not, then the hypothesis about a random process is not a reasonable one, because a deduction made from it, namely that the data ought to follow a Poisson distribution, is false.

Suppose, as a second example, that a series of experiments has been carried out at a research centre to determine the yield of strawberry plants when given a complete fertiliser. An experimental unit consisted of four plants of a standard variety; let us assume that the average yield per unit was 3 kg and that the variation in yield figures among the units followed a normal distribution. So in examining the results of a new experiment on the same variety, with units of four plants and the same fertiliser but carried out elsewhere, we may feel that the yield ought to be the same, and therefore set up the hypothesis that our resulting yield in this new experiment is normally distributed with mean 3. In practice (see Chapter 15) it would be more common to have *several* different fertiliser treatments in the experiment, each treatment applied to several different units, and we would postulate that the average yields under all the treatments were *the same*, without specifying any actual value. Thus in order to compare two of these treatments, we require to examine the difference between the means of two samples from normal populations, setting up the hypothesis that this difference is 0.

Obviously hypotheses cannot be set up on purely statistical grounds, but will embody any biological knowledge available at the outset of an experiment, or before data are collected. (One modern school of thought among statisticians assumes that at the outset a probability distribution can be given for the parameters in the hypothesis being tested, and that this probability distribution is modified in the light of the experimental results. Since the specification of these *prior probabilities* is often difficult, even arbitrary, in biological work, we shall not pursue these ideas in this book.) In any event, no experimenter or statistician would waste time testing statistically the reasonableness of a hypothesis which would be indefensible biologically if accepted. We shall see shortly that an element of uncertainty must be attached to acceptance or rejection; therefore we can never *prove* any particular hypothesis *correct*, but only show that it is (or is not) a *reasonable explanation* of the data available.

Any set of data gathered by observation or experiment inevitably constitutes a small sample from a large population, and so is subject to the random variation described in earlier chapters. Clearly, therefore, we cannot expect such a set of data to conform *exactly*, in the numerical sense, to any hypothesis, however well thought out. Hence we can examine only whether a set of data conforms to a hypothesis *sufficiently closely*, in a statistical sense to be defined; and if it does so, we accept the hypothesis as reasonable not only for that particular set of data but for the general population from which it was drawn (cf. Chapter 4). Even at this stage, assuming conformity to hypothesis, it is quite possible for someone else to put forward another hypothesis, perhaps a fundamentally different one, which may also be compatible with the data; when this happens, eventual choice between the two hypotheses must be made on biological grounds. After testing conformity, the statistician cannot make any further objective contribution to the argument. It should be noted, however, that if a new hypothesis is constructed *on the basis* of the available data, rather than before seeing them, it cannot command the same credence; there is no guarantee whatever that a repeat of the experiment will give data conforming to the new hypothesis, for this might have been fundamentally influenced by a particular accidental pattern present in the set on which it was based. For example, one of the units receiving a particular treatment may have suffered some damage that was not detected.

There is of course nothing to prevent a new hypothesis, formed by studying the results of previous experiments, from being tested in a later experiment. Indeed most biological problems need a whole research and development programme, not simply one or two experiments. A complete programme will include not only tests of significance—hypothesis tests—but the use of confidence intervals (Chapter 9).

8.2 The basic test using the normal distribution

The simplest test of significance, or hypothesis test, will now be described in detail, not because others are less important but because it illustrates well the logical steps followed in any test. Also, many other tests can be reduced to this one, which uses the normal distribution in standard form, $N(0, 1)$, described in Chapter 7.

8.2.1 Test

A single observation is given, and its numerical value is z. The hypothesis under test, called the *Null Hypothesis*, is that z has been picked at random from $N(0, 1)$.

On this hypothesis, the *possible* values of z are unrestricted, since a normal variate may take any value whatever, positive or negative, large or small. But we have already remarked in Chapter 7 that as we proceed some distance from the mean, in either direction, the frequencies in the normal distribution become very small indeed. From Table VII, we were able to find the values of z, illustrated in Fig. 7.3, which cut off quite small parts of the lower tail of $N(0, 1)$: $2\frac{1}{2}\%$ of this distribution lies below $-1\cdot96$, $\frac{1}{2}\%$ lies below $-2\cdot58$, and $\frac{1}{20}\%$ lies below $-3\cdot29$. By the symmetry of the normal curve, we can at once say that $2\frac{1}{2}\%$ of $N(0, 1)$ lies above $+1\cdot96$, $\frac{1}{2}\%$ above $+2\cdot58$, and $\frac{1}{20}\%$ above $+3\cdot29$. The central 95% of $N(0, 1)$ therefore lies in the range $(-1\cdot96, +1\cdot96)$, the central 99% in $(-2\cdot58, +2\cdot58)$, and the central $99\cdot9\%$ in $(-3\cdot29, +3\cdot29)$.

In other words, when choosing a single member at random from $N(0, 1)$ we have a probability of 0·025 that it will have a value numerically less than $-1·96$, a probability of 0·025 that it will be greater than $+1·96$, and a probability of 0·95 that it will lie in the range $(-1·96, +1·96)$.

Therefore if we set up the Null Hypothesis that the observation we take is from the $N(0, 1)$ distribution, we can claim that it *should* lie in $(-1·96, +1·96)$; however, this does not *have* to happen—there is merely a large probability, 0·95, that it will. *If the Null Hypothesis is true,* $P(-1·96 < z < +1·96) = 0·95$. (The symbol $|z|$, 'modulus of z', is useful: this stands for its numerical value *ignoring sign*, and so the statement just made can be written $P(|z| < 1·96) = 0·95$. Likewise, $P(|z| > 1·96) = 0·05$ on the Null Hypothesis.)

We can continue sampling many members from $N(0, 1)$, and every time (*before* noting the numerical value) we can make the claim that z will lie in the range $(-1·96, +1·96)$. In the long run we shall expect to be wrong 5% of times; because even when we are sampling from $N(0, 1)$ 5% of the sample members should have values outside $(-1·96, +1·96)$.

But when only one member is sampled, which we suppose is from $N(0, 1)$, and its value lies outside this range, what can we say? Either we have picked an *unlikely* member, one of the sort that we expect with probability only 0·05; or our N.H. is incorrect, so that we are not in fact selecting from a $N(0, 1)$ population and z does not therefore have to lie in the range $(-1·96$ to $+1·96)$.

When carrying out a significance test, we make the rule that if $|z| > 1·96$, (i.e. $z > +1·96$ or $z < -1·96$), the N.H. that z comes from $N(0, 1)$ is rejected, and the value of z is called *significant at the 5% level. Thus the essential step of logic in a significance test is to assume that when an unlikely value arises it is the N.H. which is at fault.* No value of z is completely impossible, so we never show conclusively that the N.H. is wrong, but only obtain more or less strong evidence to suggest that it might be. If the value of $|z| > 1·96$, the evidence is fairly strong; however when $|z| > 2·58$ the evidence is stronger, and when $|z| > 3·29$ it is stronger still, because if the N.H. is true, values of $|z| > 2·58$ should occur with probability only 0·01 (1%), and $|z| > 3·29$ with probability only 0·001 ($\frac{1}{10}$%).

We have had to give a specific technical meaning to *unlikely*: an 'unlikely' value is one which has small probability—how small we shall discuss further below. And the word *significant* is really short for 'significant evidence against the Null Hypothesis'. Sometimes the phrase 'highly significant' is attached to values outside $(-2·58, +2·58)$, that is those in the extreme 1% of the distribution, and 'very highly significant' to values outside $(-3·29, +3·29)$, in the extreme 0·1%.

Summary of test

Null Hypothesis: the given value of z has been picked at random from $N(0, 1)$. Test: inspect the value of z, and if it is less than $-1·96$ or greater than $+1·96$ (i.e. if $|z| > 1·96$) reject the N.H. at the 5% level of significance. If $|z| > 2·58$, reject the N.H. at the 1% level, and if $|z| > 3·29$ reject the N.H. at the 0·1% level. (Other values of z are given in Table I).

8.2.2 Choice of levels of significance

The values $P = 0·05$, 0·01 and 0·001 used above are the commonly used levels of significance; we are not willing to reject a N.H. when the value of z being examined is more likely than once-in-twenty (i.e. probability 5%) to have arisen if the N.H. were true. It is also useful to have the stricter standards of unlikeliness, once-in-100 (1%) and once-in-1000 ($\frac{1}{10}$%) to be applied when it would be particularly undesirable to reject a N.H. that was, in fact, true. (It is often suggested that, when comparing two sets of data by the method to be described in Test 8.7.3, only those differences which appear significant at 5% are worth following up in subsequent experiments, those which are significant at 1% can be safely reported in publications, and those which are significant at 0·1% can be regarded as soundly established—always assuming that they make biological sense.)

Some areas of biological work by their nature demand rather different treatment of 'significance', an important example being in testing new drugs where a Null Hypothesis will often be that a new drug is no better than an old one. To reject the N.H. means we change to the new drug: how much we want to do this depends on the present performance of the old drug as well as the evidence in favour of the new one. More specialised medical texts should be consulted to carry this discussion further.

8.3 A general normal distribution, $N(\mu, \sigma^2)$

In practice, no data are likely to have mean 0 and variance 1; but the ideas of Section 8.2 are used very widely indeed in statistics after standardisation of other distributions. When the normal distribution is a satisfactory model for a set of data, the mean μ and the variance σ^2 may be specified in two ways: in industry, we may have a target size μ to which a manufacturing process should be working, and the maximum variability that is allowed will give an upper limit to σ^2. In biological and agricultural work, by contrast, we usually have to rely on past ('historical') data to say what μ and σ^2 have been under suitable practical conditions, and we shall compare new data against this 'standard'. Example 8.3.6 below illustrates this.

8.3.1 Test
A single value x is given, from a normal population whose variance is known to be σ^2; test the hyopothesis that the mean of this population is equal to some specified value μ. The test is to calculate $(x-\mu)/\sigma = z$, and examine the numerical value of this in the same way as for z in Test 8.2.1. We are employing here the property described in Chapter 7, that if x is $N(\mu, \sigma^2)$ then $(x-\mu)/\sigma$ is $N(0, 1)$. So if z is not an acceptable member of $N(0, 1)$, we can equally well say that x is not an acceptable member of $N(\mu, \sigma^2)$.

8.3.2 Example
Suppose that a Null Hypothesis has been set up which states that an observation x is distributed according to $N(7·25, 1·69)$. A value $x = 3.35$ is then observed; is this observation consistent with the hypothesis?

Calculate

$$z = \frac{x - \mu}{\sigma} = \frac{3 \cdot 35 - 7 \cdot 25}{\sqrt{1 \cdot 69}} = -\frac{3 \cdot 90}{1 \cdot 30} = -3 \cdot 0.$$

This lies between those z values required for significance at 1% and 0.1%, which are respectively 2·58 and 3·29; it is thus less likely than once-in-100 to arise by chance if the N.H. is true (though not so unlikely as once-in-1000) and we may reject the N.H. at the

1% level of significance (though not at 0·1%). The observed value of x is described as *significant at 1%*.

8.3.3 Example

A Null Hypothesis states that x should be $N(4\cdot50, 0\cdot36)$; an observed value of x is 5·57. Thus

$$z = \frac{5\cdot57 - 4\cdot50}{\sqrt{0\cdot36}} = \frac{1\cdot07}{0\cdot60} = 1\cdot78$$

which is less than the value of z required for significance at the 5% level (namely 1·96) and so in this case we shall accept the N.H.

8.3.4 Example

A Null Hypothesis states that x should be $N(0\cdot40, 0\cdot09)$, and a value $x = -0\cdot80$ is observed. This time,

$$z = \frac{-0\cdot80 - 0\cdot40}{\sqrt{0\cdot09}} = -\frac{1\cdot20}{0\cdot30} = -4\cdot0,$$

greater than $-3\cdot29$ in numerical value so that the N.H. is rejected at the 0·1% level of significance.

8.3.5 Example

A N.H. states that x is $N(15\cdot09, 1\cdot44)$, and an observation $x = 17\cdot43$ is made. Therefore

$$z = \frac{17\cdot43 - 15\cdot09}{\sqrt{1\cdot44}} = \frac{2\cdot34}{1\cdot20} = 1\cdot95,$$

a value which is just on the borderline of significance at the 5% level. Now this does not provide strong evidence for rejecting the N.H., but neither does it support its acceptance with any great confidence. In such a case, the statistical test has not made any very definite contribution towards our attitude to the N.H. (except perhaps to prevent us from accepting it unreservedly) and before it can do so we must ask for more data: the test will then be based on a value of $\bar{x}$, as in Test 8.4.1 (Section 8.4), and not just on a single value of x.

8.3.6 Example

A cytologist has studied chromosome sizes in a large number of healthy persons, and found that for one particular chromosome the ratio of its long arm to its short arm is normally distributed with mean value 1·75 and variance 0·0025. He measures the same ratio (long arm/short arm) for the same chromosome in a patient suspected to have some genetic abnormalities; the value of the ratio measured is 1·61. Shall he classify the patient as healthy or not?

Test 8.3.1 applies here, for we are given a value, $x = 1\cdot61$; and the N.H. states that in a healthy patient x should be taken from the distribution $N(1\cdot75, 0\cdot0025)$. Thus

$$z = \frac{1 \cdot 61 - 1 \cdot 75}{0 \cdot 05} = -\frac{0 \cdot 14}{0 \cdot 05} = -2 \cdot 80.$$

This is greater in numerical value than $-2 \cdot 58$, and so is significant at 1%; at the 1% level, therefore, we may reject the N.H. that x was taken from $N(1 \cdot 75, 0 \cdot 0025)$, and we are at liberty to say that this patient cannot reasonably be classed as healthy. Of course, it is quite possible that in spite of the statistical evidence, the cytologist will not want to classify a patient on the basis of one single observation, and he would take more chromosome samples and calculate more arm ratios; the mean value of these ratios would then be tested by Test 8.4.1. Alternatively he would measure other possible indicator characteristics as well as the arm ratio.

8.4 The mean of a sample from a normal distribution

Few people would be happy to make decisions, of the type we have been discussing so far in this chapter, on the basis of a single observation; the existence of natural variation makes this a dangerous, even meaningless, pastime. The mean of a reasonable number of observations is a much more satisfactory basis, particularly as this allows the scatter in the observations to be studied as well as their location. In Section 7.7 we stated a result which is of very great use in applied statistics: *when a random sample of n observations is drawn from $N(\mu, \sigma^2)$, the sample mean will follow the distribution $N(\mu, \sigma^2/n)$.*

8.4.1 Test

A random sample of n observations $x_1, x_2, \ldots , x_n$ is given from a normal population whose variance is known to be σ^2. Test the Null Hypothesis that the mean of this distribution is μ.

The observed value $\bar{x}$ is distributed as $N(\mu, \sigma^2/n)$ and therefore $z = \frac{(\bar{x}-\mu)}{\sqrt{\sigma^2/n}}$ is $N(0, 1)$ if the N.H. is true.

8.4.2 Example

A sample of eight observations, drawn at random from a normal distribution whose variance is known to be 9, has a mean of $5 \cdot 75$. Test the N.H. that the mean of this distribution from which the sample was drawn had the value 4.

The mean of eight observations from $N(4, 9)$ will itself have distribution $N(4, \frac{9}{8})$, i.e. $N(4, 1 \cdot 125)$. So if the N.H. is true, the sample mean $5 \cdot 75$ will have this distribution $N(4, 1 \cdot 125)$. Thus

$$z = \frac{5 \cdot 75 - 4 \cdot 00}{\sqrt{1 \cdot 125}} = \frac{1 \cdot 75}{1 \cdot 06} = 1 \cdot 65$$

which is not significant at the 5% level (being less than $1 \cdot 96$), so that we can accept the sample mean as a member of $N(4, 1 \cdot 125)$ and therefore we can accept the N.H. that the mean of the distribution from which the sample was drawn was 4.

8.4.3 Example

After incubation for 24 hours at 18 °C, spores of a particular species of fungus are examined under a microscope. Over a long period of study the average length of their germ-tubes has proved to be 8·2 scale-divisions, and the variance of length 0·052. A new incubator is installed, and spores of the same species incubated in it for 24 hours at 18 °C. A random sample of 20 of these spores is selected, and examined in the customary way under the microscope. The mean of the 20 germ-tube lengths is 8·32 scale-divisions. Test the N.H. that the growth rate over the 24 hours is unchanged.

Test 8.4.1 may be applied, on the N.H. that the growth rate is still $N(8·2, 0·052)$. The mean of 20 observations from $N(8·2, 0·052)$ will then follow the distribution $N(8·2, 0·052/20)$, and we therefore test whether $\bar{x} = 8·32$ is an acceptable member of this latter distribution.

$$\frac{\bar{x} - \mu}{\sqrt{\sigma^2/n}} = \frac{8·32 - 8·20}{\sqrt{0·052/20}} = \frac{0·12}{\sqrt{0·0026}} = \frac{0·12}{0·051} = 2·35$$

which is significant at the 5% level. We therefore reject the N.H. at the 5% significance level, and consider that we have evidence against the theory that growth rate is unchanged.

8.4.4 Example. Mean of a large sample

Applying the result stated in Section 7.7, we can use this test for the mean of any large sample, whether it came from a normal distribution or not. The result is only approximate, but in samples from distributions that are not too skew the approximation will be good.

Suppose that 360 plants of a food crop in a plantation took a mean of 83·3 days after planting out to reach maturity. The variance of these records was 15·87, but experience suggests that the times to maturity will not exactly follow a normal distribution. However, since the sample is so large, we can say that the mean time $\bar{x}$ will be approximately $N(\mu, \sigma^2/n)$, and we can use $s^2 = 15·87$ as a very good estimate of σ^2 in such a large sample. In a different strain of this crop, which has usually been the one grown in this area, mean maturity time has been 82·5 days. Is there evidence of a change?

We must test the Null Hypothesis that there is no change, i.e. $\mu = 82·5$. Therefore $\bar{x} = 83·3$, $\mu = 82·5$, and by using $\sigma^2 = 15·87$ we can standardise the distribution of $\bar{x}$ and obtain

$$z = \frac{83·3 - 82·5}{\sqrt{\frac{15·87}{360}}} = \frac{0·8}{\sqrt{0·04408}} = \frac{0·8}{0·21} = 3·81.$$

This is a very unlikely value, in the extreme 0·1% tail of the $N(0, 1)$ distribution, and so the N.H. is rejected. It appears that the new strain is different from the old. (Note that because we did not have an A.H. which told us which direction any change was likely to be in, we have carried out a two-tail test.) (See Section 8.5.)

8.5 Alternative hypothesis: one- or two-tailed test?

In all the examples so far, we have used the idea suggested by Fig. 7.3(b), that *both* extremities of the normal curve (z greater than $+1·96$ or less than $-1·96$) contain *unlikely* values of z, whose appearance in the result of a significance test makes us doubt the validity of the N.H. But if we do reject the N.H., what hypothesis shall we accept? Until now, we have tacitly accepted as *Alternative Hypothesis* (or A.H.) in these cases that if our observed values or means do not come from a normal distribution with the specified value of μ, then they come from a distribution with some other value of μ which is not specified at all. This unspecified μ can equally well be higher or lower than that stated in the N.H. We shall refer to this A.H. as the *vague* one; and in all these cases we have rejected extreme values in *either* direction as in Fig. 7.3(b), so that the tests so far described may be called *two-tailed* ones because both ends (or *tails*) of the normal curve are rejected.

But sometimes the A.H. is not just the vague one: it actually specifies something about the alternative value of μ which is to be accepted if the value of μ in the N.H. is rejected. The A.H. may set a numerical value to μ, or it may say that the alternative value of μ is greater than that in the N.H. without actually giving a numerical value (or, of course, that the alternative value is less than that in the N.H.). The first of these possibilities is included in the other two (and the third is dealt with in a way very obviously similar to the second) so that we can illustrate the general method for one-tailed tests by a single example.

8.5.1 Example

A pathologist knows that mycelial colonies (*spots*) of a common species of fungus, when incubated under standard laboratory conditions for a fixed period of time, grow to an average diameter of 26 units, and that the distribution of diameters in a large population of mycelial spots is $N(26, 16)$. A closely related species, visually similar, has colony diameter distributed as $N(30, 16)$. In a given sample of 50 spots, the mean diameter proves to be 29·4 units; the null hypothesis is that the sample is from $N(26, 16)$, but if not then it must be from $N(30, 16)$.

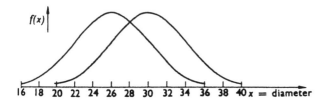

Fig. 8.1   Distributions of colony diameters, $N(26, 16)$ and $N(30, 16)$ (Example 8.5.1).

Figure 8.1 shows the two possible normal distributions for this example, the one to the left being the N.H. upon which calculations are based: on this, the mean of 50 observations will be $N(26, \frac{16}{50})$, and so we test 29·4 as a possible member of $N(26, \frac{16}{50})$. This gives

$$z = \frac{29\cdot4 - 26\cdot0}{\sqrt{0\cdot32}} = +\frac{3\cdot40}{0\cdot57} = +6\cdot0$$

which is extremely large, and so is significant at 0·1%. Therefore we reject the N.H. and automatically accept the A.H. This is reasonable, here, because the A.H. specifies a larger mean than the N.H. and this sample did have a larger mean. But what if it had had a mean considerably smaller than 26? The A.H. under those circumstances would have been more unreasonable than the N.H., and could not sensibly have been accepted: in fact, *all* negative values of z are more likely to occur on the N.H. than on

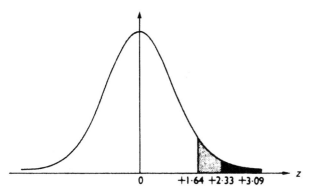

Fig. 8.2   Values of z for one-tailed significance tests at 5%, 1%, 0.1% levels.

the A.H., so we cannot in this situation reject the left-hand tail. All the region of rejection of values of the mean calculation on the N.H. must be at the *right-hand* end, as shown in Fig. 8.2; we can only reject those values that can be better explained by the A.H. So if we wish to make a test at the 5% significance level, we must use that value of z which will exclude 5% of the area under the curve at the *upper end alone*, namely $z = +1 \cdot 64$. (This is of course the numerical value of z, taken both positively and negatively, which in a two-tailed test would exclude 10% of the total area; and so it appears as a 10% entry in most standard tables.) The corresponding values of z for one-tailed tests at the 1% and 0.1% levels are $2 \cdot 33$ and $3 \cdot 09$ respectively.

A Null Hypothesis will only be rejected in favour of an Alternative if *both* of these conditions are met:
(1) the observed data are unlikely to have arisen on the N.H.,
(2) the A.H. provides a better explanation of the data.

8.6 Differences between two means

In Section 7.8 result (b), we noted that the difference between two normally distributed measurements X and Y was also normally distributed, $N(\mu_1 - \mu_2, \sigma_1^2 + \sigma_2^2)$; and in Section 7.7 we saw that when a random sample of n observations is taken from $N(\mu, \sigma^2)$ its mean $\bar{X}$ will follow $N(\mu, \sigma^2/n)$. Combining these two results, we obtain a very valuable way of comparing two sets of data.

When a random sample of n_1 items is taken from the population X which is $N(\mu_1, \sigma_1^2)$ and a random sample of n_2 items taken from the population Y which is $N(\mu_2, \sigma_2^2)$ then the difference in means $\bar{X} - \bar{Y}$ is

$$N\left(\mu_1 - \mu_2, \ \frac{\sigma_1^2}{n_1} + \frac{\sigma_2^2}{n_2}\right).$$

In significance testing we shall proceed by standardising an observed difference $\bar{x} - \bar{y}$ to obtain

$$z = \frac{(\bar{x} - \bar{y}) - (\mu_1 - \mu_2)}{\sqrt{\frac{\sigma_1^2}{n_1} + \frac{\sigma_2^2}{n_2}}} \quad \text{which is } N(0, \ 1).$$

8.6.1 Example

The manufacturer mentioned in Exercise 7.10.7 instals a second machine which produces items whose mass follows a normal distribution. On the basis of 125 items from this machine, the mean mass is 454·62 kg and the variance of mass is 25·95 kg^2. Is there a difference between the means of the production from the two machines?

Label the masses of the product from the first machine (Exercise 7.10.7) X, and from the second machine Y. Then $\bar{X} = 455\cdot51$, $\bar{Y} = 454\cdot62$, $\sigma_1^2 = 33\cdot33$, $\sigma_2^2 = 25\cdot95$, $n_1 = 100$, $n_2 = 125$. Since both distributions of mass are normal, the difference $\bar{X} - \bar{Y}$ will be

$$N\left(\mu_1 - \mu_2, \frac{33\cdot33}{100} + \frac{25\cdot95}{125}\right).$$

In order to test whether there is a difference, of unspecified size, we have to set up a Null Hypothesis that in fact there is *no* difference. We do not know which machine is expected to produce heavier items, so a two-tail test will be appropriate.

On the N.H., $\mu_1 = \mu_2$; the observed difference $\bar{X} - \bar{Y}$ will then be $N(0, 0\cdot3333 + 0\cdot2076) = N(0, 0\cdot5409)$. The observed difference is $455\cdot51 - 454\cdot62 = 0\cdot89$. Hence

$$z = \frac{0\cdot89 - 0}{\sqrt{0\cdot5409}} = \frac{0\cdot89}{0\cdot735} = 1\cdot21.$$

This value of z, tested as $N(0, 1)$, gives no evidence of any difference, because it lies within the central 95% of that distribution; we cannot therefore reject the N.H. that $\mu_1 = \mu_2$.

8.6.2 Example. Approximate normality—means of 'large' samples·

We also noted in Section 7.7 that provided the sample size, n, was large the distribution of $\bar{X}$ would be *approximately* normal, $N(\mu, \sigma^2/n)$, even if the distribution of X itself was not normal. The definition of 'large' really depends on the shape of the X distribution; if this is not far from being symmetrical (as we may be able to check by some of the methods in Chapter 2) then n need not be very large for the approximation to be good; $n = 50$ might be sufficient. The less symmetrical X is the larger n needs to be; but for most naturally occurring distributions that we are likely to meet, a sample of $n = 250$ would be more than enough to apply this approximation. We could also use the sample variance as a very good estimate of σ^2, based on such a large sample of observations.

The weights of a root crop from individual plants are measured on two sites. We want to test a Null Hypothesis that the mean weights on the two sites are the same: μ_1 (site A) $= \mu_2$ (site B). On site A, the total of 150 plants is 468·0 kg, with estimated variance 1·84 kg^2 and on site B the total of 120 plants is 391·2 kg, with estimated variance 1·36 kg^2. We will use the estimated variances from these large samples and so $\sigma_1^2 = 1\cdot84$, $\sigma_2^2 = 1\cdot36$, giving $\sigma_1^2/n_1 = \frac{1\cdot84}{150} = 0\cdot012\ 267$ as the variance of the mean weight on site A and $\sigma_2^2/n_2 = \frac{1\cdot36}{120} = 0\cdot011\ 333$ as the variance of the mean weight on site B. The observed means are $\bar{x} = \frac{468\cdot0}{150} = 3\cdot12$ on site A and $\bar{y} = \frac{391\cdot2}{120} = 3\cdot26$ on site B. Because the samples are both large, we shall not need to examine the individual weights further (e.g. by drawing a histogram) but will apply the normal approximation to the distribution of $\bar{X} - \bar{Y}$. It is $N(\mu_1 - \mu_2, \sigma_1^2/n_1 + \sigma_2^2/n_2)$ which is $N(0, 0\cdot012\ 267 + 0\cdot011\ 333)$ on the N.H., i.e. $N(0, 0\cdot0236)$. The observed difference is $\bar{x} - \bar{y} = 3\cdot12 - 3\cdot26 = -0\cdot14$, and so

$$z = \frac{-0 \cdot 14 - 0}{\sqrt{0 \cdot 0236}} = -\frac{0 \cdot 14}{0 \cdot 1536} = -0 \cdot 911$$

which is clearly not significant as $N(0, 1)$ so we cannot reject the N.H.: we shall say that $\mu_1 = \mu_2$ is a satisfactory explanation of the data.

8.7 Testing means when variances are not known

So far we have assumed that whenever we study the possible values of means in normally distributed data we will already know the variances. If samples are large, so that we can use the variance in a sample as though it were the variance in the parent population, there is no difficulty. There is one situation that demands different treatment: *if small samples of data (n<30, say) are taken from normal distributions whose variances are unknown, we cannot use the statistic z to examine values of means.*

The basic reason for this is that $z = (\bar{x} - \mu)/\sqrt{\sigma^2/n}$ is normally distributed, but if σ^2 is replaced by s^2 then $t = (\bar{x} - \mu)/\sqrt{s^2/n}$ is no longer normal. The sample variance $s^2 = \sum(x_i - \bar{x})^2/(n-1)$ has to be used because σ^2 is not known. Instead of putting σ^2, which is a given number, into z we have to use s^2, which must be recalculated afresh for each new sample that is taken, in t. We cannot expect that t, which contains *two* estimated values $\bar{x}$ and s^2 calculated from the sample, will have the same distribution as z, which contains only $\bar{x}$. However, as we shall see below, the distribution of t approaches $N(0, 1)$ for larger sample sizes; in fact t approximates $N(0, 1)$ very closely for $n > 30$, and so this new problem really only arises for quite small n. Nevertheless, samples of less than 30 observations are very likely to arise in biological work, and indeed in several other areas of applied statistics, and therefore t does need serious study.

We are still assuming that the original data $x_1, x_2, \ldots, x_n$ are a random sample from a normal distribution. If small samples of data are available from non-normal distributions, t does *not* apply and the methods of Chapter 10 will be required.

W. S. Gosset (who wrote under the pen-name *Student*) studied the distribution of t and found that it had a symmetrical frequency curve, bell-shaped like the normal, with mean 0 and a variance which depended on the size n of the observed sample. In fact the parameter of the t-distribution is not exactly n, but is the divisor $(n-1)$ in the expression for s^2: this value $(n-1)$ is called the *degrees of freedom* (d.f.) of s^2 and of t, and the expression $(\bar{x} - \mu)/\sqrt{s^2/n}$ is said to be *distributed as t with* $(n-1)$ *degrees of freedom*, often abbreviated to $t_{(n-1)}$. Those values of $t_{(n-1)}$ which exclude the extreme 5% or 1% or 0.1% of the area under its frequency curve can be found, to be applied in significance testing in exactly the same way as we have already used z; but because the exact shape of the t-distribution depends on its d.f., a t-table (Table I) needs to contain a row of these values for each of the d.f. 1, 2, 3, 4, ... Since t is one of the many distributions which converge to normality, in the manner discussed in Chapter 7, most tables quote values of t up to 30 d.f. only, t being very close to $N(0, 1)$ for degrees of freedom above 30. The shapes of the curves of $t_{(6)}, t_{(20)}$ and $N(0, 1)$ are compared in Fig. 8.3.

We shall use the same methods as previously for examining means, but the values of z have to be replaced by those for $t_{(n-1)}$. So if a random sample consisting of 16 observations is available, we shall be using $t_{(15)}$ instead of z in tests. From Table I, we shall need to replace 1·96, 2·58, 3·29 for 5%, 1%, 0.1% significance by the t-values

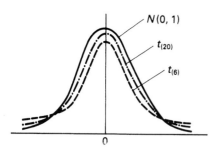

Fig. 8.3   Comparative shapes of the distributions $t_{(6)}$, $t_{(20)}$, $N(0, 1)$.

2·131, 2·947, 4·073. Table I contains, in the last row against the '∞' symbol under d.f., the values of z for the commonly used probabilities, both for one- and two-tail tests. This should help to make sure we use the right numbers when applying the t-distribution: we simply have to identify the correct column by looking at the z-values in the last row, and then move up to the row with the required d.f.

8.7.1 Test

A random sample of n observations is given, from a normal distribution of unknown variance, and the sample mean $\bar{x}$ and variance s^2 are calculated. To examine the N.H. that the true mean of the normal distribution is μ, test $t_{(n-1)} = (\bar{x} - \mu)/\sqrt{s^2/n}$ as t with $(n-1)$ degrees of freedom.

8.7.2 Example. One sample: testing the mean

Test whether the following random sample of 10 observations could reasonably have been taken from a normal distribution whose mean was 0·6: [1·2, 2·4, 1·3, 0·0, 1·0, 1·8, 0·8, 4·6, 1·4, 1·3].

The sum of all 10 observations is 15·8, and so $\bar{x} = 15\cdot8/10 = 1\cdot58$. Now $\sum_{i=1}^{10} (x_i - \bar{x})^2$, where x_i denotes a typical one of the 10 observations, is calculated by using the equivalent formula (Section 3.6) $\sum_{i=1}^{10} x_i^2 - \left(\sum_{i=1}^{10} x_i \right)^2 / 10$. The sum of the squares of the x_i is

$$(1\cdot2)^2 + (2\cdot4)^2 + \cdots + (1\cdot3)^2 = 38\cdot58$$

and

$$\frac{\left(\sum_{i=1}^{10} x_i \right)^2}{10} = \frac{(15\cdot8)^2}{10} = \frac{249\cdot64}{10} = 24\cdot964.$$

So

$$\sum_{i=1}^{10} (x_i - \bar{x})^2 = 38\cdot58 - 24\cdot964 = 13\cdot616.$$

The sample variance s^2 is

$$\frac{1}{9} \sum_{i=1}^{10} (x_i - \bar{x})^2 = \frac{13\cdot616}{9} = 1\cdot513.$$

Therefore test

$$\frac{\bar{x} - \mu}{\sqrt{s^2/n}} = \frac{1\cdot58 - 0\cdot60}{\sqrt{1\cdot513/10}}$$

as $t_{(n-1)}$, i.e. $t_{(9)}$, which gives

$$t_{(9)} = \frac{0\cdot98}{\sqrt{0\cdot1513}} = \frac{0\cdot98}{0\cdot389} = 2\cdot52.$$

On consulting Table I, we see that the 5% point for $t_{(9)}$ is 2·26, or in other words that the central 95% of the t distribution with 9 d.f. lies between $-2\cdot26$ and $+2\cdot26$. The observed value 2·52 is greater than the figure in the Table, i.e. is outside the central 95% of the distribution; we therefore reject the Null Hypothesis at the 5% significance level: we *cannot* reasonably suppose that the sample was drawn from a distribution whose true mean was 0·6.

In the above example, no definite Alternative Hypothesis was specified, but had this been done we should have had to consider whether to carry out a two-tailed test as above, or a one-tailed test. If a one-tailed test were decided on, it would be performed in exactly the same manner as the illustration of Example 8.5.1; the 10%, 2% and 0·2% entries in Table I are employed, in the same way as described for z.

We compared two means from normally distributed samples of data in Example 8.6.1. There we had large samples of data and used the sample variances as σ_1^2, σ_2^2. Now we consider the very common problem illustrated in Fig. 8.1, namely that of comparing two normal samples where the location is different *but the variance is the same*. There will be a hypothesis about the difference $(\mu_1 - \mu_2)$, but the variances in the two populations from which the samples are drawn are both equal to σ^2, the value of which is unknown and so has to be estimated from the data. This condition on the variance is necessary for the *t*-distribution to apply; if there were different values σ_1^2, σ_2^2 which we had to estimate from the samples the problem would be more complicated and we will not study that in this book. However, there are sometimes ways of obtaining an approximate result, which we note in Example 8.7.5 below.

When the results of a designed experiment are being analysed, as in Chapter 15, much the most common need is to see whether $\mu_1 = \mu_2$, and thus to test whether $(\bar{x} - \bar{y})$ could be 0; the value of σ^2 is very unlikely to be known, but has to be estimated from the information provided by all the treatments in the experiment. In other applications, $\mu_1 - \mu_2$ may not be 0.

8.7.3 Test. Difference between two means: variance not known

Two samples are given, both drawn from normal populations having the same variance (which is unknown); their sizes are n_1, n_2 respectively and their means $\bar{x}, \bar{y}$. Test the N.H. that both samples were drawn from the same population, i.e. had the same mean as well as the same variance.

The variance of $\bar{x} - \bar{y}$ is $\sigma^2/n_1 + \sigma^2/n_2$, estimated by $s^2/n_1 + s^2/n_2$. First calculate a pooled variance, that is a variance based on both samples,

$$s^2 = \frac{1}{n_1 + n_2 - 2}\left(\sum_{i=1}^{n_1}(x_i - \bar{x})^2 + \sum_{j=1}^{n_2}(y_j - \bar{y})^2\right).$$

Now test

$$\frac{\bar{x} - \bar{y}}{\sqrt{s^2(1/n_1 + 1/n_2)}} \quad \text{as } t_{(n_1+n_2-2)}.$$

Notice that when we estimated a variance from a single sample, we took the sum of squares of deviations $\sum(x_i - \bar{x})^2$ and divided it by the degrees of freedom $(n-1)$. When we want to make an estimate based on two samples, we *add* the sums of squares of deviations for *both* samples, $\sum(x_i - \bar{x})^2 + \sum(y_j - \bar{y})^2$, and divide by the sum of d.f., $(n_1-1)+(n_2-1)$. This gives an estimate s^2 whose d.f. are $(n_1-1)+(n_2-1)$, i.e. n_1+n_2-2. (We could extend this method to use three or more samples in the same way.) A good way of remembering this is

$$\frac{\text{'sum of sums of squares'}}{\text{'sum of degrees of freedom'}}.$$

When we have s^2, the variance of $(\bar{x} - \bar{y})$ is estimated as $s^2/n_1 + s^2/n_2$.

NOTE: We should always check that s_1^2 and s_2^2, the variances in the two samples, are not significantly different before we pool them. The F-test in Chapter 11 is needed for this.

8.7.4 Example

Seven plants of wheat grown in pots and given a standard fertiliser treatment yield respectively 8·4, 4·5, 3·8, 6·1, 4·7, 11·2 and 9·6 g dry weight of seed. A further eight plants from the same source are grown in similar conditions but with a different fertiliser and yield respectively 11·6, 7·5, 10·4, 8·4, 13·0, 7·0, 9·6, 13·2 g. Test whether the two fertiliser treatments have different effects on seed production.

For the first sample $\bar{x} = 6\cdot900$ and $\sum_{i=1}^{7} (x_i - \bar{x})^2 = 48\cdot88$; and for the second, $\bar{y} = 10\cdot0875$ and $\sum_{j=1}^{8} (y_j - \bar{y})^2 = 39\cdot8685$. Note that we do not estimate the variances of x and y separately, but use only these sums of squares in the pooled estimate

$$s^2 = \frac{48\cdot8800 + 39\cdot8685}{13} = 6\cdot8268.$$

The value of $(n_1 + n_2 - 2) = 7 + 8 - 2 = 13$ is the d.f. for t, the samples are assumed to be normally distributed (we can reasonably assume this for measurements such as crop yield), and the hypothetical value for $(\bar{x} - \bar{y})$ is 0.

So we have the test

$$t_{(13)} = \frac{\bar{x} - \bar{y} - 0}{\sqrt{s^2(\frac{1}{7} + \frac{1}{8})}} = \frac{6\cdot900 - 10\cdot0875}{\sqrt{(\frac{15}{56}) \times 6\cdot8268}} = \frac{-3\cdot188}{\sqrt{1\cdot8286}} = \frac{-3\cdot188}{1\cdot35} = -2\cdot36.$$

On consulting Table I we find that this value is significant at 5%, so casting considerable doubt on the N.H. that the two original distributions, from which the two samples were taken, had the same mean. This in turn leads us to reject, at the 5% level, the hypothesis that the two fertiliser treatments had the same effect.

8.7.5 Example. Possibly unequal variances

(a) If neither of n_1, n_2 is very small, say both are at least 20, *and* they are approximately the same size (e.g. 20 and 24) then the pooled value of s^2 can be used as in Test 8.7.3 and the resulting 't' can be tested as $t_{(n_1 + n_2 - 2)}$ to a good approximation.

(b) If the variances in the two samples, s_1^2 and s_2^2, are *not* the same, and a pooled s^2 is not calculated but instead $t = (\bar{x} - \bar{y})/\sqrt{s_1^2/n_1 + s_2^2/n_2}$ is used, we may test it *approximately* as $t_{(n_1 + n_2 - 2)}$. This test serves for smaller sample sizes than quoted in (a) above; however there is *no* simple satisfactory test when n_1 and n_2 are both very small and very unequal (e.g. $n_1 = 4$ and $n_2 = 9$). We now illustrate (b).

Two samples from normal distributions gave, respectively, means of 4·29, 4·18, and variances of 6·89, 2·55, based on 10, 12 observations. Can these samples be from distributions with the same mean?

Test

$$\frac{4\cdot29 - 4\cdot18}{\sqrt{6\cdot89/10 + 2\cdot55/12}} \text{ as } t_{(20)}$$

i.e.

$$t_{(20)} = \frac{0\cdot11}{\sqrt{0\cdot6890 + 0\cdot2125}} = \frac{0\cdot11}{\sqrt{0\cdot9015}} = \frac{0\cdot11}{0\cdot95} = 0\cdot12$$

which is certainly not significant, so that we can accept equality of means even though the test is an approximate one.

8.7.6 Paired data in the *t*-test

The two-sample *t* test, for comparing two means, assumes independent sets of data, obtained from different items all randomly selected from the available population of material. Unless this is so, the test is not valid because the theory upon which it is based does not apply.

Sometimes data consist of measurements at successive times taken on the *same* units, e.g. the growth of an organism over a fixed period, or the blood pressure of a patient before and after receiving a drug treatment. When several like organisms or patients are studied, there will be systematic differences between them: some patients will have higher blood pressure than others naturally, although the difference between the blood pressures 'before' and 'after' will more likely have similar values throughout the population.

Such data must be analysed as *one* sample of *differences*. Otherwise the systematic component of variation between patients is still present, as well as the drug effect that we want to study.

If these differences, removing the patient-to-patient variation, can be treated as a sample from a normal distribution, a single-sample *t*-test will be appropriate. Even if there is quite a noticeable variation between patients, so that their individual records may not be normally distributed (or even approximately so), these differences between before-and-after readings are much more likely to be sufficiently nearly normal for the test to be satisfactory.

8.7.6.1 Example

Bean seedlings were grown in pairs of adjacent pots, one pot of each pair being watered twice as often as the other (with half as much water each time). Heights in 12 such pairs were:

Pair	I	II	III	IV	V	VI	VII	VIII	IX	X	XI	XII	
more regular	34·2	26·4	22·8	29·4	22·8	27·6	23·6	30·5	28·3	32·9	24·4	31·3	
less regular	28·2	25·1	19·7	22·6	24·3	23·4	26·9	27·8	28·3	26·7	24·3	31·7	Total
Difference	6·0	1·3	3·1	6·8	−1·5	4·2	−3·3	2·7	0·0	6·2	0·1	−0·4	25·2

The sum of squares of the differences is 170·22; the mean difference is 2·10; and the estimated variance of difference is 10·6636.

The N.H. in this situation will often be 'true mean difference $= 0$'. Then

$$t_{(11)} = \frac{2 \cdot 10 - 0}{\sqrt{\frac{10 \cdot 6636}{12}}} = \frac{2 \cdot 10}{0 \cdot 943} = 2 \cdot 23,$$

which just reaches significance in a two-tail test at the 5% level. This provides some evidence that there is a difference due to regularity of watering.

If, wrongly, the two-sample test had been carried out, the pair-to-pair variation would have been so large that no difference would have been detected.

8.8 Approximate normality

As explained in Chapter 7 many variates are *approximately normal* in the sense that under suitable conditions on the parameters their distributions will be very close indeed to normality in shape even though not exactly normal. If we use tests based on normal theory for these measurements, we obtain very good indications of whether or not to accept such Null Hypotheses as are set up; but in so far as the distributions are not exactly normal, the significance levels applied will not be exactly 5%, 1% or 0·1% but only approximately so. Therefore we are bound to refer to this class of test as *approximate*, in contrast to the *exact* tests when data really are normal, although unless we are very careless about the conditions under which the tests are applied the approximation will be a very close one.

We noted in Chapter 7 that much of the importance of the normal distribution derives from the Central Limit theorem. Under the conditions of that theorem we may apply Test 8.4.1 to the *means* of samples drawn at random from *any* distribution, provided these samples are *large*. We have already done this in Example 8.4.4, and in Section 8.6 for differences; we shall restate it here and then apply it to measurements following the Poisson distribution.

For example, if $\bar{x}$ is the mean of $n = 500$ observations from any distribution, whose mean $\mu = 10$ and variance $\sigma^2 = 50$ are given, the sample is such a large one that the distribution of $\bar{x}$ will be very close indeed to $N(10, \frac{50}{500})$, i.e. $N(10, 0·1)$, and this can be used to test the significance of a given numerical value of $\bar{x}$. It is wise to require $n > 100$ before applying this approximation, though if the distribution from which the sample is drawn is itself not too far from normality, the method will work well enough for smaller samples.

8.8.1 Poisson distributions
8.8.1.1 Example. A single Poisson observation
The average rate of emission of radioactive particles from a source was measured over a long period, and found to be 10 particles per unit time. After an experimental

treatment had been applied to the source, a further sample was examined and emitted 17 particles in unit time. Test the N.H. that the rate of emission is unchanged.

The variate r, the number of emissions per unit time, follows the Poisson distribution, but since its average is greater than 5 we may obtain a close approximation to its distribution by employing the normal with the same mean and variance, i.e. $N(10, 10)$. Thus the new observation, 17, has to be tested as a single member of $N(10, 10)$ in the manner shown in Section 8.3.

We have

$$z = \frac{17 - 10}{\sqrt{10}} = \frac{7 \cdot 00}{3 \cdot 16} = 2 \cdot 21$$

which is certainly significant at the 5% level, and so the N.H. is rejected: it is *not* reasonable to say that the emission rate is unaltered.

8.8.1.2 Example. The mean of Poisson data

Given a random sample of n observations from a Poisson distribution whose mean is μ, the sample mean $\bar{r}$ will be approximately normally distributed, $N(\mu, \mu/n)$. If μ is small, the sample size n will need to be fairly large; but for $\mu > 5$ the Poisson is already approximately normal and so n may be quite small.

The mean background radiation in a laboratory is at the rate of six particles per hour. A plant grown with a labelled nutrient emits 7·8 particles per hour on average, when observed for a four-hour period. Is there evidence of an increase in radioactivity when the plant is present?

Applying the normal approximation, the four-hour mean $\bar{r}$ will follow $N(6, \frac{6}{4})$, i.e. $N(6, 1\cdot5)$. Thus $z = (7\cdot8 - 6\cdot0)/\sqrt{1\cdot5}$ is approximately $N(0, 1)$; $z = \frac{1\cdot8}{1\cdot225} = 1\cdot47$, which is well within the central 95% of the standard normal distribution and therefore provides no evidence for rejecting a Null Hypothesis that the rate has not altered due to the plant.

8.8.1.3 Example. The difference between two means

The number of insects caught in a light trap per unit time is recorded at a research site on a dark night; the total for 10 recording periods is 65 insects. When the study is repeated on a moonlight night, the total for eight recording periods is 43. Test whether there is any evidence that the numbers caught in darkness and moonlight are different.

The mean in darkness, $\bar{r}_1$, is 6·5 and its distribution can be approximated by $N(6\cdot5, \frac{6\cdot5}{10})$. Similarly $\bar{r}_2$, the mean in moonlight, is 5·375 and may be approximated by $N(5\cdot375, \frac{5\cdot375}{8})$. The difference $\bar{r}_1 - \bar{r}_2$ is therefore approximately normally distributed with variance $0\cdot65 + 0\cdot672 = 1\cdot322$. The Null Hypothesis to be tested is that there is no difference in mean rates, and so the observed difference is to be tested as $N(0, 1\cdot322)$. The actual value of the difference is $6\cdot5 - 5\cdot375 = 1\cdot125$, and $z = (1\cdot125 - 0)/\sqrt{1\cdot322} = \frac{1\cdot125}{1\cdot15} = 0\cdot98$, which as a standardised $N(0, 1)$ observation provides no evidence of any difference in mean catch. It is well in the central 95% of the standard normal distribution whatever test we use; but we might wonder whether a one-tail or a two-tail test is the more appropriate. Unless we can argue convincingly that the correct Alternative Hypothesis implies that the change in mean can only be one way—either it *must* increase or it *must* decrease—the wise thing is always to do a two-tail test. So unless the entomologist carrying out the study had been quite sure that a one-tail test should be done, and could justify this in the report of the study, the statistician should advise that a two-tail test be done.

8.8.2 Binomially distributed data and proportions

In a binomial distribution, the mean is not itself the most important piece of information, although (since we know n) we will often use the fact that the binomial mean is equal to np to find an estimate of p. It is the population value p, and its estimate from a sample consisting of n members, which will be most interesting. This is where the normal approximation is useful: from Section 7.7, when n is fairly large and p is not too near 0 or 1, the number r of a special type in a sample of n members is approximately distributed as $N(np, np(1-p))$. The sample estimate, $\hat{p}$, of p is found as r/n and therefore can be approximated as $N(p, p(1-p)/n)$.

8.8.2.1 Example

A plant breeder knows that if the skin colour of a tomato produced by crossing two given parents is determined by a single factor, three-quarters of a batch of seedlings will be red and the remainder striped. Upon growing 100 such seedlings, he observes that 64 of them are red and 36 striped. Can the hypothesis of a single factor be accepted?

This requires a test of whether the observations on the sample of $n=100$ seedlings are consistent with the hypothesis that $p=\frac{3}{4}$ in the population. In this problem, we are told that $p=\frac{3}{4}$, and so may use this value of p in the formal statement of the Null Hypothesis, which now reads 'the observed value $\hat{p}$ is a member of $N(\frac{3}{4}, \frac{3}{4} \times \frac{1}{4} \div 100)$, i.e. $N(\frac{3}{4}, \frac{3}{1600})$'.

The observed $\hat{p}=r/n=64/100=0.64$. Therefore we test as $N(0, 1)$

$$z = \frac{\hat{p} - \frac{3}{4}}{\sqrt{\frac{3}{1600}}} = \frac{0.64 - 0.75}{0.043} = \frac{-0.11}{0.043} = -2.56,$$

significant at 5% (almost at 1%), which leads us to reject the N.H. that $p=\frac{3}{4}$, and so to look for a less simple genetic law to explain the results.

As in all other significance tests, either a one-tail or a two-tail test may be required according to the Alternative Hypothesis (which in this example is simply 'p is not equal to $\frac{3}{4}$').

8.8.2.2 Example. Differences between proportions

The variance of an estimated proportion $\hat{p}$ is $p(1-p)/n$ which must itself be estimated from the sample as $\hat{p}(1-\hat{p})/n$. Given two independent samples of data, the variance of the difference between the estimated proportions $\hat{p}_1$ and $\hat{p}_2$ will be $\frac{\hat{p}_1(1-\hat{p}_1)}{n_1} + \frac{\hat{p}_2(1-\hat{p}_2)}{n_2}$. Relying again on the normal approximation, so long as n_1, n_2 are fairly large and p_1, p_2 are not too near 0 or 1, the observed difference $\hat{p}_1 - \hat{p}_2$ is approximately normally distributed with mean $(p_1 - p_2)$ and variance $\frac{p_1(1-p_1)}{n_1} + \frac{p_2(1-p_2)}{n_2}$ which is estimated from the data as $\frac{\hat{p}_1(1-\hat{p}_1)}{n_1} + \frac{\hat{p}_2(1-\hat{p}_2)}{n_2}$.

Two laboratories are examining methods of sterilising equipment. Laboratory A reports that 332 out of 480 items treated in a standard way gave sterile samples in a bacteriological test. Laboratory B, using the same method, reports 335 sterile out of 420. Do their results appear to be the same?

$$\hat{p}_1 = \frac{332}{480} = 0.6917; \quad (1 - \hat{p}_1) = 0.3083; \quad n_1 = 480.$$

$$\hat{p}_2 = \frac{335}{420} = 0.7976; \quad (1 - \hat{p}_2) = 0.2024; \quad n_2 = 420.$$

The observed difference $\hat{p}_1 - \hat{p}_2 = -0.1059$; the Null Hypothesis will be that the true difference $p_1 - p_2 = 0$. The variance of the difference is estimated as

$$\frac{0.6917 \times 0.3083}{480} + \frac{0.7976 \times 0.2024}{420} = 0.000\,444\,273 + 0.000\,384\,367 = 0.000\,828\,64,$$

and its square root is 0.0288. Therefore $z = \frac{(-0.1059 - 0)}{0.0288}$ is approximately $N(0, 1)$; its value is -3.68, which is significant at 0.1%. The evidence points very strongly to different results from the two laboratories. [NOTE. There is an alternative way of doing this test, using the chi-squared distribution (Section 12.4).]

8.9 MINITAB

Two sets of data may be compared both by using graphical methods and by calculating summary measures. We have already noted how to do these things for a single sample, but when two samples are to be compared we want to make the presentation as helpful as possible so that the salient features of the comparison are clear.

Dotplots of the two sets of data can be placed underneath one another, but will be on different measurement scales unless the command SAME is added to the instructions. If that is done, the same measurement scale is used for both, and location and scatter are much easier to compare: we can see which (if either) is more spread out, and with careful scrutiny we can also assess whether either or both may be skew. All of these things, together with sample size, determine which significance tests may properly be used.

SAME may also be used with HISTOGRAM and STEM-AND-LEAF, although it is not quite so easy to use for comparison purposes the layouts that are produced in these cases. Finally, the DESCRIBE command can be added, and will give all the information listed in Section 3.9 on two rows directly beneath one another, one for each data set.

For *paired* data we need to number the pairs of observations 1, 2, 3, . . . and then by using MPLOT the pairs of data can be displayed on the same graph. Those of the original set (e.g. patients before treatment) are labelled A and those of the second (after treatment) B. The display will put 1, 2, 3, . . . on the horizontal axis and A, B will be plotted against these on the vertical scale; the actual letters A, B will appear unless both measurements have the same value, in which case 2 will be printed. Thus we can see, for example, whether A is always greater than B, and whether the differences are about the same size in each pair. DESCRIBE can also be applied to the differences (B − A). Again these items of information help us to decide whether a 'paired' test will be valid.

These methods can be extended to three or more data sets and will be useful in looking at the results of designed experiments (Chapter 15 onwards).

Significance tests for Null Hypotheses about the values of means can also be called by the commands ZTEST or TTEST. When σ^2 is known, ZTEST requires its value to be given as well as the value of μ on the Null Hypothesis. The data are given in a column and so n is determined within the routine in MINITAB. The output gives the value of z together with the probability of it occurring on the N.H. (the *P-value*). We can then at once assess its 'significance' without needing to use tables: if the probability is less than 0.05 the result is significant at 5%, and so on. If we forget to give a value for μ, ZTEST assumes that it is zero. When a one-tail test is needed then ALTERNATIVE must be included in the instructions, with a coded value equal to -1 if μ in the Alternative Hypothesis is less than that in the N.H. and $+1$ if the A.H. value of μ is greater than that in the N.H.

When σ^2 is not known, TTEST is used instead. Once the data have been listed in a column, n will be known and s^2 is calculated within the routine, which outputs the t-value and its probability of occurrence on the Null Hypothesis.

If the normal approximation is to be used for tests on measurements that follow binomial distributions, including proportions, or Poisson distributions the necessary instructions to construct mean and variance have to be added to the routine ZTEST.

For differences between means when samples of data are independent, another routine will be needed, either TWOSAMPLE-T or TWOT according to the way in which data are stored. Contrary to the recommendations made in most Applied Statistics textbooks, this does not automatically use a pooled variance (as in Example 8.7.3) unless it is told to do so by including the additional command POOLED; without this it will produce the equivalent of Example 8.7.5. The command ALTERNATIVE may also be included with the same coding (-1 or $+1$) as in ZTEST. The two sets of data are given to the routine as input, and the output includes their means and standard deviations as well as the results of the test. Note that a separate test for two large samples is not needed because these tests without the extra command POOLED cover this case.

8.10 Exercises on Chapter 8

8.10.1 In each case (a)–(f) test the Null Hypothesis that the given observation x came from the specified normal distribution, against the Alternative Hypothesis as stated.

(a) $x = 4.34$. N.H. is that x is $N(2.75, 1.64)$. A.H. is that μ is not equal to 2.75.
(b) $x = 11.28$. N.H. is that x is $N(15.3, 2.72)$. A.H. is that μ is not equal to 15.3.
(c) $x = -1.03$. N.H. is that x is $N(-0.55, 0.096)$. A.H. is that μ is not equal to -0.55.
(d) $x = 21.92$. N.H. is that x is $N(18.7, 3.24)$. A.H. is that μ is greater than 18.7.
(e) $x = 1.45$. N.H. is that x is $N(3.0, 0.35)$. A.H. is that μ is less than 3.0.
(f) $x = 0.80$. N.H. is that x is $N(0, 0.1)$. A.H. is that μ is less than 0.

8.10.2 Sixty-four observations are selected at random from a normal distribution whose variance is 25. Their mean is calculated and found to be 11.1. Test the hypothesis that the true value of the population mean is 10.

8.10.3 Twenty guinea-pigs from specially bred laboratory stock are fed on a standard diet for a fortnight from birth. Their mean increase in weight is 28 units. In the past, the mean for a very large number of similar animals has been 29.8 units, and the variance of weight-increase 25 units. It is thought that these 20 animals may be different because they have been housed in a newly-installed pen. Test whether their weight increase does differ from the result obtained for the large population.

8.10.4   Two hundred observations are selected at random from a distribution whose mean is thought to be 5 and variance known to be 8. The mean of the 200 observations is 4·77; test whether the hypothesis, that the population mean is 5, is acceptable.

8.10.5   Five hundred seeds of a variety of cabbage, supplied by a particular seedsman, contained 425 that germinated when sown. It is claimed that this variety achieves 90% germination. Do this seedsman's seeds appear to satisfy this claim?

8.10.6   The numbers of isopods in each of 50 sampling quadrats were recorded, and found to follow a Poisson distribution; the mean number of isopods per quadrat was 2·2. Test the hypothesis that the true mean, over the whole area from which the samples were drawn, is 2·0. Justify your method.

8.10.7   A group of 10 strawberry plants is grown in ground treated with a chemical soil conditioner, and the mean yield per plant is 114 g. Experience has shown that when the same variety of strawberry is grown under similar conditions, but with no soil conditioner, the mean has been 110 g and the variance 84. Test whether it can reasonably be claimed that the soil conditioner had a beneficial effect on yield.

8.10.8   A standard drug used to relieve pain after a certain type of operation has been successful with 80% of patients. A new drug is tried, in the belief that it will be more successful, on 200 patients, and 170 of these gain relief. Is the belief in the new drug justified?
    What could be said if (a) 152 or (b) 140 of the patients had gained relief?

8.10.9   The weights of a large number of children in the same age-group in a region have been measured, and the mean and variance in this population are respectively 102 lb and 49 lb^2. If 100 children from an adjacent region are weighed and their mean weight is 99 lb, is there any evidence that the population in the second region differs from that in the first? If only five children had been available, how would you have examined their weight records?

8.10.10   A random sample of 10 observations of animals' growth under standard conditions gave 96·7, 84·3, 101·8, 78·3, 110·6, 93·4, 87·8, 91·3, 98·2, 88·7 cm. Test the hypothesis that the data came from a distribution with mean value 100 cm.

8.10.11   Eight tobacco leaves were used in an experiment to compare the damaging effects of two preparations A and B. Half of each leaf was treated with A and the other half with B; the number of lesions on each half leaf at the end of the experiment was recorded, as given below. The difference $d =$ (number of lesions caused by A − number caused by B) is also given. Test the hypothesis that A, B cause the same amount of damage.

Leaf number	(1)	(2)	(3)	(4)	(5)	(6)	(7)	(8)
Preparation A	31	20	18	17	9	8	10	7
Preparation B	18	17	14	11	10	7	5	6
Difference d	13	3	4	6	−1	1	5	1

What assumptions have had to be made in the analysis, and do they seem to be justified on looking at the data?

8.10.12   The following table shows the number of leaf pairs per shoot on *mercurialis perennis* after a fixed period of growth. Calculate mean, median, variance and standard deviation for these figures. Would you prefer mean or median as a measure of central tendency?

Number of leaf pairs	1	2	3	4	5	6	7	8	9
Number of shoots	0	2	5	19	60	57	32	3	2

Test the hypothesis that the true mean number of leaf pairs at this stage of growth is 5.

8.10.13   Windfall apples were examined for insect attack. Of 60 windfalls in an orchard, 32 had been attacked, while of 105 apples harvested from nearby trees 41 had been attacked. Does this indicate that the proportion of apples attacked is the same in windfalls as in harvested fruit?

8.10.14   In an experiment to determine the effect of hormones in male chicks, day old chicks from the same batch were randomly assigned to two groups. One group received androsterone (A) and the other dehydroandrosterone (C). The chicks were caged and fed under identical conditions. The comb sizes measured after two weeks were:

A: 57, 120, 101, 137, 119, 117, 104, 73, 53, 68, 118 units
C: 89, 30, 82, 50, 39, 22, 57, 32, 96, 31, 88 units

Is there evidence that the choice of compound affects comb size?

8.10.15   Eight observations were selected at random from a normal distibution, and their values were 1·6, −0·8, 0·1, −0·4, 1·2, 0·7, 0·3, 0·5. Test the hypothesis that the normal distribution had mean 0·1.

8.10.16   Two samples of observations on the diameters of mycelial colonies produced the following results. Sample A, of 11 observations, gave a mean value of 6·65 and a variance of 15·2824; sample B, of 16 observations, gave a mean value of 4·28 and a variance of 8·0275. Find a pooled estimate of variance, and use it in testing whether the samples could have come from distributions with the same mean. (Assume that the distributions of diameters are normal.)

8.10.17   The average distance travelled by one constituent of a dye on a paper chromatograph is known accurately to be 6·0 cm. The distance travelled by a second constituent of the dye was measured on a sample of 12 paper chromatographs. The mean distance travelled was found to be 5·75 cm, and the sample variance was 0.44. Is there any evidence to suggest that the mean distance travelled by the second constituent differs from that of the first?

9 Confidence intervals for means

9.1 Estimating parameters

In the early stages of a research programme, it may be possible to suggest a suitable distribution for the statistical model of our measurements, but little or nothing will yet be known about the values of parameters, such as mean and variance, in this distribution. In this situation significance tests are not possible; estimates of the parameters must be made before any specific hypotheses can be set up. The mean of a sample, from any distribution whatever, is a good estimate of the true population mean; the sample proportion $\hat{p}$ is a good estimate of the population p; the sample variance (using $(n-1)$ as divisor) is a good estimate of the true population variance. These are 'point estimates', giving a numerical value for the mean which we know will change from one sample to the next. They do not give any idea how *precisely*, μ, p or σ^2 is determined by the available sample of data.

An **interval estimate** provides a range, or interval, that has a high probability (95%, 99%, sometimes 99.9%) of containing the true population value. Note that it is the interval which is a random variable, depending on the sample values x_1, x_2, . . . ,x_n observed: the parameter value, though at present unknown, is fixed.

If we did know μ and σ^2 in a normal distribution, we could say for a single observation 'x lies between $\mu - 1.96\sigma$ and $\mu + 1.96\sigma$', or for the mean of n observations '$\bar{x}$ lies between $\mu - 1.96\sigma/\sqrt{n}$ and $\mu + 1.96\sigma/\sqrt{n}$'. Both these statements would be true *with probability 0.95*.

In symbols,

$$P(\mu - 1.96\sigma/\sqrt{n} < \bar{x} < \mu + 1.96\sigma/\sqrt{n}) = 0.95.$$

We can make an algebraic rearrangement of the inequality in brackets without altering the probability of it being true. For present purposes, assume that σ^2 is known: we return to this point later. A sample of data will give a value for $\bar{x}$. Suppose we do *not* actually know μ. We can gain information on it by rewriting the inequality as

$$P\left(\bar{x} - 1.96\frac{\sigma}{\sqrt{n}} < \mu < \bar{x} + 1.96\frac{\sigma}{\sqrt{n}}\right) = 0.95$$

and using the numerical values of σ (known), n (sample size) and $\bar{x}$ (sample mean). This gives a *95% confidence interval for* μ. In other words, this interval has probability 0.95 of containing the true value of μ.

9.1.1 Example

In the analysis of nitrogen content of the leaves of plants, a laboratory uses a method which gives results whose standard deviation is 0·025 ppm. The results follow a normal distribution whose mean depends on what experimental treatment the plants have received. A random sample of 15 plants from one treatment gives a sample mean of 2·426 ppm.

Using the method explained above, we compute

$$\bar{x} - \frac{1\cdot96\sigma}{\sqrt{n}} = 2\cdot426 - \frac{1\cdot96 \times 0\cdot025}{\sqrt{15}} = 2\cdot426 - 0\cdot013 = 2\cdot413$$

and

$$\bar{x} + \frac{1\cdot96\sigma}{\sqrt{n}} = 2\cdot426 + 0\cdot013 = 2\cdot439.$$

With probability 0·95 ('with 95% confidence') we may claim that the true mean μ lies in the interval (2·413 to 2·439). This is *a 95% confidence interval for μ*.

If we wanted to have a higher probability that the interval contined μ, we could replace 1·96 by 2·58 to obtain a 99% confidence interval, or by 3·29 for a 99·9% confidence interval. For the 99% confidence interval, $\bar{x} - (2\cdot58\sigma/\sqrt{n}) = 2\cdot426 - 0\cdot017$, giving 2·409 as the lower limit to the interval; the upper limit is $2\cdot426 + 0\cdot017 = 2\cdot443$, and so the 99% confidence interval may be stated as (2·409 to 2·443).

Using 3·29 instead of 1·96, the limits of the 99·9% confidence interval are $\bar{x} - (3\cdot29 \times 0\cdot025)/\sqrt{15} = 2\cdot426 - 0\cdot021 = 2\cdot405$, and $2\cdot426 + 0\cdot021 = 2\cdot447$. Although we can never be absolutely sure we have located the true mean on the basis of one sample, we can be 99·9% confident that the interval (2·405 to 2·447) does contain μ.

9.1.2 Example. Difference between two means

The formula for a 95% confidence interval for μ can be stated in words as *sample mean plus or minus 1·96 times its standard error*, since the variance of $\bar{x}$ is σ^2/n (Section 7.7). This can be extended to the case of the difference $(\mu_1 - \mu_2)$ between two means from normal distributions where the variances are known. The statement in words now is *difference between sample means plus or minus 1·96 times the standard error of this difference*. The variance of the difference is $\sigma_1^2/n_1 + \sigma_2^2/n_2$, in our usual notation.

Suppose that the laboratory in Example 9.1.1 has two machines which will carry out the analyses of plant samples, and that the second machine gives results which have a standard deviation of 0·032. This machine is used to analyse data from an experiment on another site, and 18 plant samples give a mean chemical content of 2·398 ppm. We want to find a 95% confidence interval for the true difference between the means on the two sites.

$$\bar{x}_1 = 2\cdot426, \ \sigma_1 = 0\cdot025, \ n_1 = 15; \ \text{and} \ \bar{x}_2 = 2\cdot398, \ \sigma_2 = 0\cdot032, \ n_2 = 18.$$

The variance of the difference is

$$\frac{\sigma_1^2}{n_1} + \frac{\sigma_2^2}{n_2} = \frac{0\cdot025^2}{15} + \frac{0\cdot032^2}{18} = 0\cdot000\ 041\ 67 + 0\cdot000\ 056\ 89 + 0\cdot000\ 098\ 56.$$

The standard error of the difference is the square root of this, namely 0·0099. The observed difference $\bar{x}_1 - \bar{x}_2 = 2\cdot426 - 2\cdot398 = 0\cdot028$, and the 95% confidence interval will be $0\cdot028 - 1\cdot96 \times 0\cdot0099 = 0\cdot028 - 0\cdot019 = 0\cdot009$ to $0\cdot028 + 0\cdot019 = 0\cdot047$.

The true difference between the mean chemical content on the first and second sites is contained in the interval (0·009 to 0·047), and we can make this statement with 95% confidence that it is correct.

9.2 Approximate normality

As in significance testing, so in estimation we very rarely know the variance of a population while we are still trying to estimate its mean. The chief value of the results quoted in Section 9.1 is that they can be applied approximately in many different problems. Results for proportions are among the most important.

9.2.1 Example

The variance of an estimated proportion p is $p(1-p)/n$. If we do not know p we have to estimate this variance as $\hat{p}(1-\hat{p})/n$. (In significance testing, we generally have a hypothesis about p which we can use in calculating the variance; in estimation obviously we do not—all we have is $\hat{p}$). Approximate 95% confidence limits to p are then

$$\left(\hat{p} - 1\cdot96\sqrt{\frac{\hat{p}(1-\hat{p})}{n}} \text{ to } \hat{p} + 1\cdot96\sqrt{\frac{\hat{p}(1-\hat{p})}{n}} \right).$$

Of 250 insects treated with a certain insecticide, 180 were killed. Set approximate 95% confidence limits to the value of p, the proportion of insects likely to be killed by this insecticide in future use.

Here $n = 250$ and $\hat{p} = \frac{180}{250} = 0\cdot720$. Thus

$$\frac{\hat{p}(1-\hat{p})}{n} = \frac{0\cdot72 \times 0\cdot28}{250} = \frac{0\cdot2016}{250} = 0\cdot000\ 806\ 4.$$

Therefore

$$1\cdot96\sqrt{\frac{\hat{p}(1-\hat{p})}{n}} = 1\cdot96\ \sqrt{0\cdot000\ 806\ 4} = 1\cdot96 \times 0\cdot0284 = 0\cdot056.$$

So approximate 95% limits for p are $(0\cdot720 - 0\cdot056 < p < 0\cdot720 + 0\cdot056)$, i.e. p lies between 0·664 and 0·776, with 95% confidence in the truth of this statement.

These are relatively wide limits for the proportion killed, and the result may not be accurate enough as a guide to future use. Expressed in terms of percentages, we are claiming a kill of somewhere between 66·4% and 77·6% when the insecticide is used in circumstances similar to those in this experiment.

Proportions need large values of n before precision becomes really good. By increasing n to 1000 we would reduce the variance to $\frac{p(1-p)}{1000}$, and supposing p is still 0·720 (i.e. 720 were killed) this is $\frac{0\cdot72 \times 0\cdot28}{1000} = 0\cdot000\ 201\ 6$. The square root, which is the standard error of the estimated proportion, is 0·0142. So the 95% confidence limits for this sample of 1000 are from $(0\cdot720 - 1\cdot96 \times 0\cdot0142) = (0\cdot720 - 0\cdot028) = 0\cdot692$ to

$(0.720 + 0.028) = 0.748$. The interval $(0.692$ to $0.748)$ is now quite narrow, and gives a good idea of the likely future values of p in the same conditions as the present data were collected. (In terms of percentage kill, if this is easier to understand, we could say that the value was between about 69 and 75.)

It is possible to obtain approximate 99% and 99·9% confidence intervals in the same way by replacing 1·96 with 2·58 (for 99%) or 3·29 (for 99·9%), which of course will make the intervals wider.

As in all normal approximations applied to proportions, p must not be too near 0 or 1 if the approximation is to be reasonably good; and the results are best when p is not too far from $\frac{1}{2}$.

9.2.2 Example. Difference between two proportions

Again for large enough n_1 and n_2, and values of p_1, p_2 not too near 0 or 1, we can apply a method similar to that in Example 9.1.2 to find confidence intervals for the true $(p_1 - p_2)$ based on sample estimates $\hat{p}_1$ and $\hat{p}_2$. The variance of $(\hat{p}_1 - \hat{p}_2)$ is $\frac{p_1(1-p_1)}{n_1} + \frac{p_2(1-p_2)}{n_2}$, which must be estimated from the samples as $\frac{\hat{p}_1(1-\hat{p}_1)}{n_1} + \frac{\hat{p}_2(1-\hat{p}_2)}{n_2}$. We may then apply the usual rule '*observed difference plus or minus 1·96 times the standard error of this difference*'.

The data of Example 8.8.2.2 have been used to test a Null Hypothesis '$p_1 = p_2$' (which we rejected), but they can also serve to give an interval estimate of the true population difference $(p_1 - p_2)$. We found that $\hat{p}_1 = 0.6917$, for laboratory A, and $\hat{p}_2 = 0.7976$, for laboratory B. The variance of the difference was 0·000 828 64, and so the standard error (the square root of this) was 0·0288.

The estimate of $(p_1 - p_2)$, the difference 'laboratory A minus laboratory B', is $0.6917 - 0.7916 = -0.1059$, and the 95% confidence interval for this difference therefore goes from $(-0.1059 - 1.96 \times 0.0288) = (-0.1059 - 0.0564) = -0.162$ to $(-0.1059 + 0.0564) = -0.050$. With 95% confidence of being correct (on the evidence of these data) we claim that the difference in proportions A *minus* B lies in the range $(-0.162$ to $-0.050)$. Stated in terms of percentages, A is less than B by between 16% and 5%.

9.2.3 Example. The mean of a Poisson distribution

There is, as we have seen, a good normal approximation to the Poisson distribution when the mean μ is at least 5: the Poisson r can be approximated as $N(\mu, \mu)$. From this, the mean $\bar{r}$ of n observations from the same Poisson distribution is approximately distributed as $N(\mu, \mu/n)$. The result allows a confidence interval to be found for μ on the basis of the sample. The observed value of the mean is $\bar{r}$ (the sample mean) and the variance of this observed value is estimated by $\bar{r}/n$. Therefore an approximate 95% confidence interval is given in the form

$$\bar{r} - 1.96 \sqrt{\frac{\bar{r}}{n}} < \mu < \bar{r} + 1.96 \sqrt{\frac{\bar{r}}{n}}.$$

As usual, the confidence level can be changed to 99% by using 2·58 instead of 1·96 and to 99·9% by using 3·29.

A sample of 80 leaves from a population of plants grown with a labelled nutrient showed an average radioactive particle count of 9·15 per 15-minute period. Set approximate 95% confidence limits to μ, the true mean rate of radioactive emission under this nutrient treatment.

The observed mean $\bar{r}$ is 9·15, and the estimate of its variance is $\frac{9·15}{80}$. The rule '*observed value plus or minus 1·96 times its standard error*' therefore gives the lower (approximate) limit as

$$\bar{r} - 1·96\sqrt{\frac{9·15}{80}} = 9·15 - 1·96\sqrt{0·114\,375} = 9·15 - 1·96 \times 0·338 = 9·15 - 0·663$$

$= 8·49$, and the corresponding upper limit as $9·15 + 0·663 = 9·81$. The approximate 95% confidence interval for μ is thus (8·49 to 9·81) particles per 15-minute period.

9.2.4 Example. Means of 'large' samples

The idea of 'approximate normality' discussed in Section 8.8 applies to means of samples from *any* distribution under suitable conditions: if the distribution of an observation x has mean μ and variance σ^2, and a large sample is taken, then $\bar{x}$ is approximately $N(\mu, \sigma^2/n)$. The sample must be very large if the x-distribution is very skew, but correspondingly smaller if x has a reasonably symmetrical distribution.

The mean height to which a sample of 75 trees have grown in a nursery in a year has been 0·682 m, and the variance was 0·1145 m². The distribution of height may not be normal but is not thought to be very skew. Set 95% confidence limits to the true value of the mean in the population of which these trees form a part.

We do not need to know the variance in the whole population, because we have the value given by a large sample (i.e. the sample variance s^2) and this will provide a very good estimate of σ^2. So an approximate 95% confidence interval for the true value of μ is

$$\left(\bar{x} - 1·96\sqrt{\frac{s^2}{n}} \text{ to } \bar{x} + 1·96\sqrt{\frac{s^2}{n}} \right).$$

Here, $\bar{x} = 0·682$, $n = 75$, $s^2 = 0·1145$ and so

$$\sqrt{\frac{s^2}{n}} = \sqrt{\frac{0.1145}{75}} = \sqrt{0·001\,527} = 0·039.$$

The lower 95% limit is therefore $0·682 - 1·96 \times 0·039 = 0·682 - 0·077 = 0·605$, and the upper limit is $0·682 + 0·077 = 0·759$, giving the interval (0·605 to 0·759) m.

9.2.5 Example. Difference between two means

We return to the data of Example 8.6.2. There we tested a Null Hypothesis that two means were equal, and found no evidence to reject it. Therefore any interval we calculate for the true difference $(\mu_1 - \mu_2)$ ought to contain the value 0: that is a very reasonable, possible value for the difference as we discovered from the significance test. We may gain further insight into the precision with which the difference has been determined by applying the same normal approximation to find an interval of the form '*observed difference plus or minus 1·96 times the standard error of this difference*'. The observed difference (site A – site B) is $3 \cdot 12 - 3 \cdot 26 = -0 \cdot 14$, and the variance of this is $\sigma_1^2/n_1 + \sigma_2^2/n_2$, in which we shall use the sample estimates and put $\sigma_1^2 = 1 \cdot 84$ and $\sigma_2^2 = 1 \cdot 36$. The estimated variance of the difference is $\frac{1 \cdot 84}{150} + \frac{1 \cdot 36}{120} = 0 \cdot 0236$ as we found in Example 8.6.2. The lower 95% confidence limit is thus

$$-0 \cdot 14 - 1 \cdot 96\sqrt{0 \cdot 0236} = -0 \cdot 14 - 1 \cdot 96 \times 0 \cdot 1536 = -0 \cdot 14 - 0 \cdot 301 = -0 \cdot 44,$$

and the upper limit is $-0 \cdot 14 + 0 \cdot 301 = 0 \cdot 16$. The 95% confidence interval for the true value of the mean difference (site A – site B) is therefore ($-0 \cdot 44$ to $+0 \cdot 16$). It does contain 0, as it must if it is to be consistent with the result of the significance test; and it has given us the additional information that the value of the difference is not very precisely determined, because the interval is rather wide relative to the actual size of the difference.

9.3 Means when the variance is unknown and the samples are small

If samples are small and are *not* taken from normal distributions, then there is no satisfactory way of constructing confidence intervals for means or differences between means. (The methods of Chapter 10, which give significance tests for this situation, cannot easily be adapted to provide intervals.) For large samples we can as usual rely on approximate normality of the means, as in Examples 9.2.4 and 9.2.5.

Student's *t*-distribution is the key to calculating confidence intervals for means of data when the samples available are small *and* we can safely assume that a normal model explains them. The theoretical results that are needed are exactly the same as for significance testing; therefore, as we saw in Section 8.7, the same methods will be used as for normal theory except that values of z (1·96 etc.) will be replaced by those of t with the appropriate degrees of freedom. Our description in words of the interval will now be '*observed mean plus or minus t times its standard error*'.

9.3.1 Example. One sample of data

In Example 8.7.2 we had a sample of 10 observations from a normal distribution and rejected at the 5% level a Null Hypothesis that the mean of that distribution is 0·6. Let us now see what the 95% confidence interval is for μ: we know it should *not* contain 0·6.

From the $n = 10$ observations, $\bar{x} = 1 \cdot 58$ and $s^2 = 1 \cdot 5129$. The t value to be used is that for $(n - 1) = 9$ degrees of freedom, and the critical values of this from Table I are 2·262 (5%), 3·250 (1%) and 4·781 (0·1%). The 95% interval is from

$$\bar{x} - t_{(n-1)}\sqrt{\frac{s^2}{n}} \quad \text{to} \quad x + t_{(n-1)}\sqrt{\frac{s^2}{n}}.$$

This is from $1 \cdot 58 - 2 \cdot 262\sqrt{1 \cdot 5129/10} = 1 \cdot 58 - 2 \cdot 262 \times 0 \cdot 389 = 1 \cdot 58 - 0 \cdot 88 = 0 \cdot 70$ to $1 \cdot 58 + 0 \cdot 88 = 2 \cdot 46$. The interval is (0·70 to 2·46).

To find a 99% confidence interval, 2·262 is replaced by 3·250 everywhere in this calculation; this will give (0·32 to 2·84). Finally, for 99·9%, the confidence interval uses 4·781 for the t value and becomes ($-0 \cdot 28$ to $+3 \cdot 44$).

9.3.2 Example. Difference between two means

This method works so long as it is correct to assume that the variance in the two populations of data is the same; as in Test 8.7.3 this is first checked by an F-test (Chapter 11). Returning to the data of Example 8.7.4, we will find a 95% confidence interval for $(\mu_1 - \mu_2)$. The data gave $\bar{x} = 6 \cdot 9000$ and $\bar{y} = 10 \cdot 0875$, the difference being $-3 \cdot 1875$. We must make the same calculation of a 'pooled' s^2 as in Test 8.7.3; the value is $s^2 = 6 \cdot 8268$ and it has 13 degrees of freedom. The 95% confidence interval is

$$(\bar{x} - \bar{y}) - t_{(n_1 + n_2 - 2)} \sqrt{s^2 \left(\frac{1}{n_1} + \frac{1}{n_2} \right)} \quad \text{to} \quad (\bar{x} - \bar{y}) - t_{(n_1 + n_2 - 2)} \sqrt{s^2 \left(\frac{1}{n_1} + \frac{1}{n_2} \right)}.$$

The sample sizes were $n_1 = 7$ and $n_2 = 8$. We found

$$\sqrt{s^2 \left(\frac{1}{n_1} + \frac{1}{n_2} \right)} = \sqrt{(6 \cdot 8268) \left(\frac{1}{7} + \frac{1}{8} \right)} = \sqrt{1 \cdot 8286} = 1 \cdot 35.$$

The 95% point of $t_{(13)}$ is $2 \cdot 160$. We now have all the information to construct the interval: it runs from $-3 \cdot 1875 - 2 \cdot 160 \times 1 \cdot 35 = -3 \cdot 1875 - 2 \cdot 916 = -6 \cdot 10$ to $-3 \cdot 1875 + 2 \cdot 916 = -0 \cdot 27$. So the interval is ($-6 \cdot 10$ to $-0 \cdot 27$).

We note again that this interval does *not* contain 0, which is correct since the significance test showed some evidence of a real difference. Also it is a very wide interval relative to the size of the observed difference, so the precision of determination is not good.

9.3.3 Example. Paired data

In this case we must work with the individual observed differences on each experimental unit, just as in Section 8.7.6, and treat these as a single sample of data from a normal distribution.

The data of Example 8.7.6.1 allow us to construct a 95% confidence interval for the true difference (once the systematic patient-to-patient difference has been removed by the pairing). We found that the observed mean difference was $2 \cdot 10$, and the variance of the differences was $10 \cdot 6636$ with 11 degrees of freedom. We rejected a Null Hypothesis that there was really no difference. The 5% point of t from Table I is $2 \cdot 201$. The method of Example 9.3.1 is required, and gives the 95% limits from $2 \cdot 10 - 2 \cdot 01 \sqrt{10 \cdot 6636 / 12} = 2 \cdot 10 - 2 \cdot 201 \times \sqrt{0 \cdot 8886} = 2 \cdot 10 - 2 \cdot 201 \times 0 \cdot 943 = 2 \cdot 10 - 2 \cdot 07 = 0 \cdot 03$ to $2 \cdot 10 + 2 \cdot 07 = 4 \cdot 17$. The interval thus is ($0 \cdot 03$ to $4 \cdot 17$). It just fails to include 0, as we expect from the result of the significance test in Example 8.7.6.1.

Note. In all these calculations of confidence intervals we use the two-tail 5% (or 1% or $0 \cdot 1$%) points of t, because we are using the argument that any value of μ that leads to a t value in the central 95% (etc.) of the appropriate t-distribution is acceptable as the parent mean for this sample of data. Since t, like z, is unbounded in its possible values this is logically the only reasonable way to construct an interval because we cannot give any definite values above and/or below which μ cannot theoretically go. In practice, in some problems μ obviously has to be positive. If in such a case a calculated interval has a negative lower limit this may well be because we have a poor, imprecise set of data so that the whole interval is actually very wide. In these circumstances it is sensible to quote the lower limit as 0, and calculate the upper limit in the usual way.

9.4 Estimation or testing?

As we remarked at the beginning of this chapter, estimation will be needed early in a research programme in order to establish values of the parameters in a statistical model (distribution) used to explain sets of data. However, when we study designed experiments, beginning in Chapter 15, we shall discover that very often it is the comparison of two means which is the basic purpose of experimentation; one may be a 'control' or 'standard' treatment currently used and the other a new, possibly improved, treatment. It does not matter exactly what the mean yield μ is under the control, and indeed this may depend on the environment in which the present experiment is being done; interest lies in whether we can alter this, and if so by how much. So it may appear that the significance-testing approach usually used, and described in Chapters 15–18, is the 'obviously correct' approach.

However, the same information can be obtained from a confidence interval calculation as from a significance test: as we have seen, if a significance test shows that two means do not appear to be different, the confidence interval for the true difference will contain the value 0. So although the method discussed later in this book is the commonly used one, it does not actually give as much information as we would get from a confidence interval. By calculating an interval, we discover how precise the experimental data are as well as what evidence they give about the one specific question asked in the significance test. We find *all* the means (or differences) which are consistent with the experimental data that we have, not simply check one value given in a hypothesis.

We shall return to this in Chapter 16 where we shall see that a statistically significant result may not be of practical importance whereas a non-significant one sometimes can be. Indeed there is a welcome trend at the present time towards using confidence intervals in preference to significance tests.

9.5 MINITAB

The command ZINTERVAL provides the 95% confidence limits $\bar{x} - (1\cdot96\sigma/\sqrt{n})$ to $\bar{x} + (1\cdot96\sigma/\sqrt{n})$ for the true value of μ in the population from which a data set $x_1, \ldots, x_n$ is drawn; the data set must be input to the routine together with the known value of σ that is to be used. If a 99%, or 99·9%, interval is required this can be asked for; the default setting is 95%.

When σ^2 is not known, TINTERVAL is used instead; this calculates s^2 and replaces σ by s in the formula. The appropriate degrees of freedom for t are used to give as output

$$\left(\bar{x} - \frac{ts}{\sqrt{n}}; \ \bar{x} + \frac{ts}{\sqrt{n}}\right).$$

For differences between means, the routine TWOSAMPLE-T (or TWOT) described in Section 8.9 provides confidence intervals as well as carrying out significance tests.

9.6 Exercises on Chapter 9

9.6.1 A random sample of 25 seeds is selected from a large population of a certain variety and planted in pots. The mean time from planting to the opening of the first leaflet is 5·85 days, with variance 4·84. Assuming that this time follows a normal distribution, test the hypothesis that its mean is 4. Find a 95% confidence interval for the mean time. Comment on the results.

9.6.2 Experience has shown that the increase in weight of guinea-pigs from a specially bred population, when fed for a standard length of time from birth on a standard diet, is normally distributed. Six newborn guinea-pigs are selected at random, and fed for the same length of time on a new diet; their weight increases are 18, 21, 12, 16, 25, 20 units. Assuming that the increase in weight remains normally distributed, set 95% confidence limits to the value of its mean.

9.6.3 Set 95% confidence limits to the proportion of patients cured by a new drug if 170 out of 200 in a clinical trial were cured. Compare with the result of the significance test in Exercise 8.10.8.

9.6.4 A random sample of 100 children had a mean weight of 99 lb. Assuming that the variance of weight is 49 lb^2 (based on a large amount of data from an adjacent part of the country), set 95% and 99% confidence limits to the true mean weight.

9.6.5 A large normal population has mean 100 and variance 10. Find 95% confidence-limits to:

(a) the value of one single observation drawn from this population;
(b) the mean of 10 randomly selected observations drawn from this population;
(c) the mean of 100 randomly selected observations drawn from this population.

If the variance had not been known, how could limits be found?

9.6.6 Fifty samples, each of one unit volume, were drawn at random from a suspension containing cells. Each sample was placed on a microscope slide and examined; it was found that the average number of cells per unit volume for these 50 samples was 4. Set approximate 95% confidence-limits to the mean number of cells per unit volume in the whole suspension.

9.6.7 Set 95% confidence limits to the true value of the difference *d* in the whole population of leaves described in Exercise 8.10.11.

9.6.8 A sample of nine plants of the same variety were grown in one type of soil in a greenhouse, and after a fixed time they were removed and dried. Their dry weights were 25·5, 22·3, 24·7, 28·1, 26·5, 19·0, 31·0, 25·3, 29·6 g. A further sample of 11 similar plants were grown in identical conditions but in another type of soil. Their dry weights were 31·8, 30·3, 26·4, 24·2, 27·8, 29·1, 25·5, 28·9, 30·0, 26·9, 29·7 g. Find a 95% confidence interval for the mean weight in each population separately. Find also an interval for the difference between the two mean weights. What assumptions have been necessary in the analysis?

9.6.9 How many patients would need to be treated with drug A mentioned in Exercise 9.6.10 below, in order to estimate the proportion of cures to within ± 3% at the 95% confidence level?

9.6.10 Two preparations of a drug, presented in the same tablet form, are tested for their efficacy in alleviating headaches. Preparation A is given to 250 patients, and 172 of them claim that it is effective; B is given to 200 patients and 158 of these claim that it is effective. Set 95% confidence limits to the true value of the difference between the proportions of patients claiming effectiveness on the two drugs.

9.6.11 Oleoresins are extracted from conifers by tapping the resin ducts in the sapwood. A modification of the usual technique is compared with the standard form of the technique using pairs of trees that are as similar as possible. Pairs from eight different species of conifer are available, and the oleoresin contents extracted in a fixed time of sampling are as follows, given as percentages of dry weight.

Species	1	2	3	4	5	6	7	8
Modified technique	1·74	0·47	1·52	0·63	0·75	5·01	1·40	0·79
Standard technique	1·57	0·54	0·93	0·57	0·68	4·90	1·72	0·75

Use a suitable method to set 95% and 99% confidence limits to the difference in content due to modifying the technique. Explain why the method is appropriate for these data.

10 Nonparametric tests for location

10.1 Assumptions needed for tests

All the significance tests described in Chapters 8, 9 and 11 depend on some assumptions about the distribution from which the sample of observations has been drawn. The χ^2 test, when used either to test the goodness-of-fit of data to a ratio-type hypothesis or to examine data arranged in a contingency table in Chapter 12, will be the only exception to this requirement. Generally we have asked that observations shall be taken from a normal distribution, or at least from one which approximates closely to normality.

When there is doubt whether the basic assumptions for the normal deviate, t- or F-tests apply, or the assumptions for the χ^2 test of variances, it may be appropriate to consider tests which do not depend on these assumptions, or indeed on any assumption at all about the form of distribution from which observations have been taken. However, in view of the Central Limit Theorem, the tests given for *sample means* in Chapter 8 apply very widely. The t-test can be used, even for quite small samples, when data have been drawn from a distribution that may not be normal, provided that the distribution is not too skew; under these conditions, a test on the sample means made using t leads to very nearly the same result as an exact one worked out specially for that distribution. We may call the t-test *robust* to departures from normality; the general definition of a robust test is one which continues to give results correct at almost the same levels of significance even when some of its basic assumptions do not hold. For a symmetrical, or nearly symmetrical, distribution we may safely use t, and so it is really only for rather skew sets of data that we should look seriously for alternative tests. Even skew data can sometimes be transformed to symmetry, as we saw in Chapter 7 where the lognormal distribution was described. In such a case, a t-test could still be used on the logarithms of the original observations.

However, tests of the *shape* of a distribution (variance and skewness) are much less *robust*, and much more sensitive to non-normality. Furthermore, if we wish to compare two samples from distributions which may be of different shape, it is useful to have a test of the difference in their locations (*centres*) which does not take shape into account at all, even to assume any similarity.

The mean and variance of a distribution are, as we have seen, always closely related to the parameters in its probability density function; tests not employing these directly are called **nonparametric**, and they generally depend on **ranking**, that is arranging a set of observations in order of size, rather than using their actual numerical values. Some of the available nonparametric tests are of high efficiency, in the sense that when the assumptions for the corresponding parametric test (e.g. the t-test when means are being examined) *are* satisfied we would not lose very much in sensitivity by applying the nonparametric one instead. So if we have any doubt about making these assumptions, we may prefer to use a nonparametric test anyway, especially if it is easy and quick to apply. Another name given to these tests is ***distribution-free***.

10.2 Location of the centre of a distribution

Corresponding to the *t*-test for a mean, there is a test of whether a given sample of data came from a distribution having median M. We use M rather than the mean, because M is more likely to be a stable measure of central tendency when the shape of a distribution is in doubt.

10.2.1 The sign test
A set of observations $x_1, x_2, \ldots, x_m$ is given. Test the N.H. that they arise from a distribution whose median value is M. The A.H. is simply 'median is not M'. Attach a sign to each one of $x_1, x_2, \ldots, x_m$, + if its numerical value is greater than M and $-$ if it is less than M. (Any x_i which is exactly equal to M must be ignored.) Denote the number of + signs by n_+, and suppose that n signs have been allocated altogether ($n = m$ if no x_i was equal to M). Now if the true value of the median is M, and we have ignored any x_i that is equal to M, the probability that any x_i will be less than M is the same as the probability that it will be greater than M, namely $\frac{1}{2}$. On this hypothesis, therefore, n_+ follows a binomial distribution whose parameters are n and $\frac{1}{2}$, and we may determine whether the observed value of n_+ lies in the extreme tail of this binomial distribution.

10.2.2 Example
A questionnaire used in a psychological assessment is thought to give a median score of 50 in a group of students doing a particular course. When tried out on 20 students of another course, it gave the scores:

$$26, 46, 39, 58, 62, 41, 65, 49, 54, 50, 61, 38, 58, 35, 27, 34, 46, 51, 29, 40.$$

Test the hypothesis that the median is 50.

We attach a ' + ' sign to any value over 50, '$-$' to any below, and ignore any exactly equal to 50. The signs are $- - - + + - + - + \cdot + - + - - - - + - -$. There are $12 -$ signs and $7 +$s, with one observation of exactly 50 which cannot be used.

If $M = 50$, individual scores are equally likely to be above or below 50, i.e. equally likely to have a $+$ or a $-$ sign. Altogether $n = 19$ signs have been allocated, and the distribution of the number of $+$s (or of $-$s) should be binomial with $n = 19$ and $p = \frac{1}{2}$. Considering $+$s, $r = 7$ is a binomial observation with $n = 19$ and $p = \frac{1}{2}$. Approximately (because p is $\frac{1}{2}$ although n is really rather too small) r is $N[np, np(1-p)]$ which is $N(19 \times \frac{1}{2}, 19 \times \frac{1}{2} \times \frac{1}{2})$ or $N(9 \cdot 5, 4 \cdot 75)$. We require the tail area $P(R \leqslant 7)$, since the observed value was 7. With continuity correction (Section 7.6), this is $P(R < 7\frac{1}{2})$. The standardised value $z = \frac{(7 \cdot 5 - 9 \cdot 5)}{\sqrt{4 \cdot 75}} = -\frac{2 \cdot 0}{2 \cdot 18} = -0 \cdot 92$, not significant and giving no evidence against the N.H. that the median is 50.

10.2.3 Example
This test can also be used in paired comparisons, where there is no numerical measurement but only a preference between two alternatives. Suppose that a tasting

panel of 25 people has compared a food product preserved in two different ways; 15 prefer method A, nine prefer method B and one has no preference. The N.H. is 'Methods A, B are equally good' and the A.H. 'there is some difference'. On the N.H., the 24 stated preferences form an observation from a binomial distribution with $n = 24$ and $p = \frac{1}{2}$, which can be approximated by $N(24 \times \frac{1}{2}, 24 \times \frac{1}{2} \times \frac{1}{2})$ or $N(12, 6)$. Since there were nine preferences for B, we may compute $P(R \leqslant 9)$ as the tail area. With continuity correction this is $P(R < 9\frac{1}{2})$. The standardised value $z = \frac{(9 \cdot 5 - 12 \cdot 0)}{\sqrt{6}} = -\frac{2 \cdot 5}{2 \cdot 449} = -1 \cdot 02$, not significant and giving no evidence of real difference between the methods in their effect on taste.

10.3 Two samples of related measurements

In some experiments, data come naturally in related pairs, rather than independent items. For example, in an experiment in which a (non-absorbent) fungicide is sprayed on to the leaves of tomato plants we may choose n leaves at random from a collection of leaves of similar age on plants growing in a greenhouse; then one surface of each leaf will be coated with fungicide and the other surface not. If we can actually count the number of developing fungal spots or lesions after a fixed period of growth in an environment infested with fungal spores, we have two parametric analyses open to us. We may carry out a randomised block analysis (Chapter 16) for two treatments (fungicide and no fungicide) in n blocks, the blocks term removing any systematic differences between the leaves. We may also calculate the difference x between the number of lesions on the untreated surface and that on the treated surface for each leaf, so giving n readings of x which we analyse as in Example 8.7.6. Taking the difference in each leaf removes between-leaf differences: we do *not* have two independent samples, each of n observations, because they arise from the *same* leaves.

So as to make these analyses, we need to assume that the lesion counts are normally distributed, or at least that the differences x are; sometimes a logarithmic or square-root transformation (Chapter 18) will help us to justify doing the analysis. But if we are not able to justify any of these methods, or if it is very difficult to count lesions, we may choose a simpler nonparametric method, illustrated by the following example which uses the same approach as Example 10.2.3.

10.3.1 Example

Forty leaves were used in a fungicide experiment, one surface of each leaf being coated with a standard (non-absorbent) fungicide, and after a week's exposure to infection the leaves were assessed as 'coated surface cleaner', 'no difference' or 'coated surface more infected'; the numbers in these categories were 28, eight and four respectively. Test whether the fungicide coating is effective.

The Sign Test 10.2.1 may again be used; $n = 32$, ignoring the 'no difference' leaves, and $n_+ = 28$, leading (with continuity correction) to $z = (27\frac{1}{2} - 16)/\sqrt{8} = 11 \cdot 5/2 \cdot 828 = 4 \cdot 07$, a value significant at the 0·1% level. There-fore we reject the Null Hypothesis of no difference.

However, we do not always want to throw away *all* the information about the sizes of differences in pairs, as the Sign Test does, and if such information is available Wilcoxon's **Matched-Pairs Signed Ranks** test can be used.

10.3.2 Test (Signed Ranks Test)

On each of m items, two records are available: x_1 and y_1 on item 1, x_2 and y_2 on item 2, and so on. The differences $w_i = x_i - y_i$ are found, and the *moduli* of these w_i are ranked in order, ignoring any w_i that are zero. The sum of ranks for the positive w_i is calculated; call this R_+. Likewise the sum of ranks for the negative w_i is called R_-; the smaller of R_+, R_- is called T. To test the Null Hypothesis that the median difference between x and y is zero, test T as an approximately normal variate with mean $\frac{1}{4}n(n+1)$ and variance $\frac{1}{24}n(n+1)(2n+1)$; this can be used for $n=10$ or more, n being m minus the number of zero ws.

10.3.3 Example

Two red food colouring agents are compared for retention of colour in various solutions after various heat treatments, with the following results.

Solution/Treatment combination	(1)	(2)	(3)	(4)	(5)	(6)	(7)	(8)	(9)	(10)
Agent A	4	4	5	1	3	3	5	6	6	3
Agent B	7	6	9	5	6	3	4	7	4	7

The records are scores on a visual 10 point scale of colour scoring. Test whether the two agents appear to differ.

If we denote score B by x and score A by y, the values of w are 3, 2, 4, 4, 3, 0, -1, 1, -2, 4 and the rank order for the moduli of w (ignoring *minus* signs) is $5\frac{1}{2}$, $3\frac{1}{2}$, 8, 8, $5\frac{1}{2}$, $-$, $1\frac{1}{2}$, $1\frac{1}{2}$, $3\frac{1}{2}$, 8. One w was zero, and does not feature in the ranking. Several cases of *tied ranks* occur; there are two values of $|w|$ equal to 1, and so two contestants for first place in the rank order. We compromise by letting them share the first two places, giving each a rank of $1\frac{1}{2}$. Similarly there are two contestants for the next place, so they must share third and fourth, with rank $3\frac{1}{2}$; two must share fifth and sixth, with rank $5\frac{1}{2}$; and the final three are *all* contenders for seventh place, so they must share seventh, eighth and ninth, with average rank $(7+8+9)/3=8$. The two negative values of w have ranks $1\frac{1}{2}$ and $3\frac{1}{2}$, so $R_- = 1\frac{1}{2}+3\frac{1}{2}=5$. The seven positive values of w have ranks which sum to $R_+ = 40$. Here $n=9$, and we can check our calculation by seeing that $R_- + R_+ = \frac{1}{2}n(n+1)=\frac{1}{2}\times 9 \times 10=45$, which it does. T is the smaller of R_-, R_+, and so is 5; we will use the normal approximation even though n is only 9 in this example. (In fact the result that it gives is very similar to that from a more exact and laborious method.)

On the Null Hypothesis that the two agents are equally good, T is, then, approximately normally distributed with mean $\frac{1}{4}n(n+1)=\frac{1}{4}\times 9 \times 10=22\cdot5$ and variance $\frac{1}{24}n(n+1)(2n+1)=\frac{1}{24}\times 9 \times 10 \times 19=71\cdot25$, so that $z=(T-22\cdot5)/\sqrt{71\cdot25}$ is approximately $N(0, 1)$. The value of z is $(5-22\cdot5)/8\cdot44 = -2\cdot07$, which is significant at the 5% level, so that we reject the Null Hypothesis.

The Signed Ranks test compares the medians of two populations, which are of similar shape; however, this shape need *not* be normal. It uses information about the size of the observations as well as ranking, and as a result it is much more efficient than the sign test; in fact it is not much less accurate than a t-test even when the assumptions for a t-test are valid. Therefore it is a very useful test for 'paired' data.

10.4 Two samples, not 'paired': the *U*-test

10.4.1 Test
This compares the medians of two independent samples, which need not contain the same number of observations; the Null Hypothesis to be tested is that the two medians are equal. The Alternative Hypothesis is that the two medians are not equal. Call the two samples I, II. Rank their members into one single order, preserving the distinction between I and II, and count the total number (U) of times that members of II follow members of I in this ranking. When neither sample size is less than 8, this count U is distributed approximately as N ($\frac{1}{2}n_1n_2, \frac{1}{12}n_1n_2(n_1 + n_2 + 1)$), n_1 and n_2 being the two sample sizes. If n_1 and n_2 are 7 or less, a special table is needed (see, for example, the book by Siegel[4]).

The logic behind the test is that on the N.H. the two samples (A, B say) should appear in random order in the joint ranking.

10.4.2 Example
The weights of leaves produced by plants from the same original source, after an experiment comparing two artificial day lengths, I and II, were:

I:	17·2	5·1	12·3	6·9	8·2	13·5	13·3	11·2	11·6	14·2	10·8	7·1		
II:	19·0	15·3	12·4	17·5	12·8	13·0	14·6	10·6	6·7	9·3	15·8	16·8	10·1	19·1

The joint ranking is:

5·1	6·7	6·9	7·1	8·2	9·3	10·1	10·6	10·8	11·2	11·6	12·3	12·4
I	II	I	I	I	II	II	II	I	I	I	I	II
12·8	13·0	13·3	13·5	14·2	14·6	15·3	15·8	16·8	17·2	17·5	19·0	19·1
II	II	I	I	I	II	II	II	II	I	II	II	II

We now count the number of times a II follows a I in this joint ranking, and call it U.

The first I has 14 IIs above it; the next three Is each have 13 IIs above; the next four Is each have 10 IIs above; the next three have seven, and the last one three. Thus $U = 14 + 13 + 13 + 13 + 10 + 10 + 10 + 10 + 7 + 7 + 7 + 3 = 117$.

$$n_1 = 12, \quad n_2 = 14, \quad \frac{1}{2}n_1n_2 = 84; \quad \frac{1}{12}n_1n_2(n_1 + n_2 + 1) = 14 \times 27 = 378.$$

Therefore the standardised value $z = \frac{(117-84)}{\sqrt{378}} = \frac{33}{19·44} = 1·70$, which is not significant in a 2-tail test. There is not sufficient evidence to reject the N.H. of equal median weights.

When two of the observed values are equal, we call this a *tie*; if there should happen to be ties between observations in the same sample, the test is not affected, but if there are ties between observations in different samples, this obviously complicates the problem of deciding how many times the observations from one sample precede those from the other in the joint ranking. This complication is solved by adjusting the formula for the variance of U; we omit details and refer readers to the text by Siegel.[4]

It can be shown that when the parent distributions are normal, U provides nearly as good a test for differences between means as does t (it is, in fact, about 95% efficient).

But unfortunately, the U-test is only satisfactory when we are comparing the medians of samples from two distributions which have the same shape. The value of the test lies in the fact that this shape need not be normal or even symmetrical; however it is not a satisfactory test for comparing, for example, the median of a symmetrical distribution with the median of a skew one, nor even the median of a slightly skew distribution with the median of a very skew one. In fact, U is less *robust* for changes in shape than is t for the effects of non-normality, and U cannot be improved by increasing the sample sizes—a most unusual situation in significance testing.

10.5 Nonparametric methods in MINITAB

The Sign Test 10.2.1 is carried out using the command STEST; the median value for the Null Hypothesis is specified, and either a one-tail or a two-tail Alternative Hypothesis can be nominated. The output gives the probability (P-VALUE) of the data observed, on the Null Hypothesis, and also the median value in the sample. There is a refinement to obtain approximate confidence intervals as well, called SINT. If data are paired (as in Example 10.2.3) the differences must first be computed and then input into STEST.

Wilcoxon's Signed Rank Test 10.3.1 uses WTEST, with specified value for the median on the Null Hypothesis and a choice between one- and two-tailed Alternative Hypotheses. The normal approximation is used in this test and so sample size should not be too small. Again there is a procedure for finding approximate intervals.

The U-test is called by the command MANNWHITNEY. The two sets of data are input, each into its own column. Output consists of n_1, n_2 and the two medians, with a test of the Null Hypothesis that the medians are equal and an interval for their true difference.

10.6 Exercises on Chapter 10

10.6.1 Twenty-five subjects undergoing a test reported the following reaction times (sec) to a stimulus:

6·6, 3·6, 5·4, 7·2, 4·7, 13·1, 2·0, 7·6, 2·3, 2·8, 15·4, 4·3, 6·7, 9·5, 11·8, 1·4, 19·7, 3·0, 7·5, 6·9, 23·3, 6·4, 14·1, 6·0, 3·8.

Test the hypothesis that the median reaction time is 7·8 sec.

Calculate the mean reaction time and comment on the result.

10.6.2 Two samples A, B, of plants of the same species growing on opposite slopes of a valley were dug up and weighed, as follows:

A (20 observations):	27·1, 40·3, 15·7, 3·9, 22·2, 36·4, 11·8, 16·3, 14·7, 16·2, 32·0, 15·7, 12·9, 27·5, 9·9, 14·4, 24·8, 7·2, 21·0, 18·8
B (12 observations):	11·7, 15·3, 19·1, 22·0, 6·7, 14·1, 19·1, 24·4, 15·8, 12·3, 28·7, 17·9

Test the hypothesis that their median weights were equal.

10.6.3 In a tasting experiment, people were presented with pairs of samples of vegetables preserved by the same method but kept for different periods of time. It was required to test the hypothesis that no difference could be detected between samples kept for a short period of time and those kept for longer. Thirty people took part, eight said there was no difference, seven preferred the longer-kept samples, and the remaining 15 preferred the shorter-kept samples. Test the hypothesis.

10.6.4   The people taking part in the experiment described in Exercise 10.6.3 were also asked to score each sample on a 10-point scale. They obtained these results:

Person	1	2	3	4	5	6	7	8	9	10	11	12	13	14	15
Short storage	3	6	5	7	8	5	4	2	7	6	4	5	8	6	5
Long storage	4	6	4	3	3	5	5	2	5	5	4	1	3	8	5

Person	16	17	18	19	20	21	22	23	24	25	26	27	28	29	30
Short storage	8	2	6	3	6	5	4	7	5	5	7	4	5	7	6
Long storage	2	5	5	3	7	3	7	7	6	4	4	1	5	6	2

Make a further statistical test, which uses the information contained in the scores, to examine the same hypothesis as previously.

10.6.5   In the experiment described in Exercises 10.6.3 and 10.6.4, no Alternative Hypothesis was offered as to which samples ought to be 'better'. A food technologist, looking at the results, claims that if there is any difference it must be the shorter-kept samples which are preferred. Examine the results (a) using only the preferences, and (b) using the scores also, in the light of this claim.

10.6.6   Two storage methods A and B for apples are compared by examining fruit for % weight loss. Pairs of similar fruits are weighed after a fixed time in store. Results for 18 pairs of Cox's Orange Pippin were:

Pair	1	2	3	4	5	6	7	8	9	10	11	12	13	14	15	16	17	18
Method A	4·6	6·5	2·9	3·5	3·8	4·2	5·1	4·7	5·2	5·2	4·1	4·3	4·7	6·0	5·2	2·6	3·5	4·8
B	5·3	6·7	3·9	3·1	3·3	5·8	5·9	4·4	4·9	6·0	4·2	4·2	5·4	6·3	6·5	2·5	3·2	4·5

Carry out (a) a suitable nonparametric test, (b) a *t*-test, to test the hypothesis that weight losses are not different under the two storage methods.

Compare the results of the two analyses.

11 Tests and estimates for variances

11.1 Studying variances

The results of experimental work are usually assessed in terms of changes to the *mean* response—crop yield, plant or animal growth, reaction time, control of pests and diseases; and there are very good reasons for this as may be clear from Chapters 7, 8 and 9. There is a great deal of theory available for studying means from a wide range of distributions, and the Central Limit Theorem validates statistical methods based on means in many situations.

Because of this we often overlook how much information the variances of sets of data can give. We can look at dot-plots, stem-and-leaf diagrams and box-and-whisker plots, and assess whether different sets of data are about equally variable or whether some are much more widely scattered than others. There could be many reasons for increased scatter. One that affects plant growth is the use of different levels of treatments such as fertilisers, where some of the treatments may allow the plants full and healthy growth while others give relatively poor responses. Those that are near the borderline between sufficiency and deficiency may give much more variable results: stronger plants are satisfactory but weaker ones may suffer quite seriously under the same treatment.

Methods for studying variances numerically and theoretically exist only for *normal* distributions; there are no approximate (or exact) results which apply widely to other distributions. Also the results which we can use are affected quite seriously by moderate departures from normality, much more so than those for means; therefore it is important to check that data are reasonably symmetrical before looking in detail at their variances.

11.2 The chi-squared distribution

If a random sample of n observations is drawn from $N(\mu, \sigma^2)$, where there is no need to know μ, we first calculate the sample variance $s^2 = \sum_{i=1}^{n}(x_i - \bar{x})^2/(n-1)$; then the expression $(n-1)s^2/\sigma^2$ follows $\chi^2_{(n-1)}$, the *chi-squared* distribution with $(n-1)$ degrees of freedom. Since this distribution is basically the sum of squares of normal variables, it is defined only for positive values. It is a highly skew distribution, and tables such as Table II need to show both upper and lower critical values for each separate degrees-of-freedom. Call the lower and upper $2\frac{1}{2}\%$ points χ^2_L, χ^2_U; Figure 11.1 shows the shape of the distribution and these points.

The expression $(n-1)s^2$ is called the *sum of squares* of the data set, because it is σ^2 times $\sum_{i=1}^{n}(x_i - \bar{x})^2$; it is a basic idea in the Analysis of Variance which we shall use

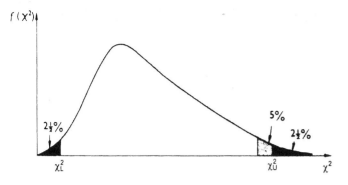

Fig. 11.1 The χ^2-distribution, showing one- and two-tailed 5% significance points. χ^2_U, χ^2_L are upper and lower $2\frac{1}{2}$% points.

frequently in Chapters 14–17. The χ^2 distribution, like t, depends on the sample size n through the parameter for degrees of freedom, which is $(n-1)$; Table II therefore has to contain a row for each d.f. The typical shape of the χ^2 curve for 3 or more d.f. is shown in Fig. 11.1; the actual position of the peak and the rates of increase and decrease of frequency on either side of it are all determined by the d.f. The hump moves steadily to the right as the d.f. increase, and the shapes of $\chi^2_{(3)}$ and $\chi^2_{(10)}$ are compared in Fig. 11.2.

There are two possible ways in which we may choose to exclude an extreme 5% part from the χ^2 distribution, illustrated in Fig. 11.1 by different degrees of shading. First, as for the normal and t distributions when used in two-tailed tests, we may choose to omit both the extreme $2\frac{1}{2}$% at the lower end and the extreme $2\frac{1}{2}$% at the upper end. Secondly, we may choose to omit only the upper 5%, accepting all values less than this, right down to 0. This second, *one-tailed* procedure is needed more often than the first, and so the standard tables of χ^2 quote, for each d.f., the values of χ^2 *above* which lie only 5%, or 1%, or 0·1% of the total area. So when we want a two-tailed procedure, we need the $2\frac{1}{2}$%, $\frac{1}{2}$% and 0·05% points, and also those points which have $2\frac{1}{2}$%, etc., below them. Now the point shown in Fig. 11.1 having $2\frac{1}{2}$% area below it can just as well be said to have $97\frac{1}{2}$% of area above it—so in Table II it appears as a $97\frac{1}{2}$% point. Example 11.3.2 illustrates how this is used.

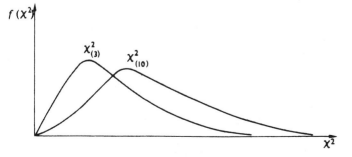

Fig. 11.2 Comparison of χ^2-distributions with 3 and 10 degrees of freedom.

11.3 Testing a hypothesis about σ^2

There are two requirements in studying variances, just as for means. First we deal with significance testing (hypothesis testing). When data are normally distributed, we may have a Null Hypothesis that says '*Variance* $= \sigma^2$', where the value of σ^2 is often based on past experience of similar measurements. The variance in a new sample of n observations is s^2, and we wish to see whether this supports or contradicts the value for σ^2 in the N.H.

11.3.1 Test

A sample of n observations is given, from a normal distribution of unknown mean; test whether the distribution could have variance σ^2. Calculate

$$s^2 = \frac{1}{n-1} \sum_{i=1}^{n} (x_i - \bar{x})^2$$

and test $(n-1)s^2/\sigma^2$ as $\chi^2_{(n-1)}$; more directly, calculate $\sum_{i=1}^{n}(x_i - \bar{x})^2$ only, and test $[\sum_{i=1}^{n}(x_i - \bar{x})^2]/\sigma^2$ as $\chi^2_{(n-1)}$. In either case accept the N.H. *variance* $= \sigma^2$ if the value of $\chi^2_{(n-1)}$ falls neither in the lower $2\frac{1}{2}\%$ nor in the upper $2\frac{1}{2}\%$ region of the χ^2 distribution.

Example 11.3.2

The sample of 11 observations 4·3, 1·8, 6·5, 3·2, 5·1, 3·9, 4·6, 4·7, 2·5, 5·8, 3·6 is thought to be taken from a normal distribution with variance 1·21. Test whether this hypothesis is reasonable.

By the usual method, we find $s^2 = \frac{1}{10}(19\cdot1764) = 1\cdot9176$; thus $19\cdot1764/1\cdot21 = 15\cdot85$ is distributed as $\chi^2_{(10)}$. Now from Table II, the upper $2\frac{1}{2}\%$ point of $\chi^2_{(10)}$ is 20·48, and the lower $2\frac{1}{2}\%$ point, i.e. the upper $97\frac{1}{2}\%$ point in the Table, is 3·25. Our calculated value, being between these, does not require rejection of the N.H., and so we suppose that the given value of the variance is reasonable.

Note 1   If, as is not common, we know the mean, μ, of the distribution, we calculate $\sum_{i=1}^{n}(x_i - \mu)^2$ as the sum of squares, and use n for degrees of freedom in χ^2.
Note 2   If we have a definite Alternative Hypothesis for σ^2, this will require us to make a one-tailed test exactly as already described for means.

11.4 Confidence limits for a variance

The second use for a sample of data from a normal distribution is to construct a 95% confidence interval for the true value of σ^2 in the population from which the sample was drawn. The logic is the same as we applied when dealing with a mean, but the calculation will be slightly different because the χ^2 distribution is not symmetrical.

We claim the value of $\chi^2_{(n-1)}$ which we can calculate for our sample should lie in the middle 95% of the distribution, i.e. between the values χ^2_L and χ^2_U (Fig. 11.1). Of course we do not know σ^2; but *any* value of σ^2 for which $(n-1)s^2/\sigma^2$ lies in the range $(\chi^2_L; \chi^2_U)$ is acceptable as a possible value of the variance in the population from which our sample came. This claim has probability 0·95 of being true when we use the lower and upper $2\frac{1}{2}\%$ values for χ^2_L and χ^2_U. So

$$\chi_L^2 \leqslant \frac{(n-1)s^2}{\sigma^2} \leqslant \chi_U^2 \text{ with probability } 0\cdot95.$$

Simple algebraic rearrangement gives the usual form of the 95% confidence interval:

$$P\left(\frac{(n-1)s^2}{\chi_U^2} \leqslant \sigma^2 \leqslant \frac{(n-1)s^2}{\chi_L^2}\right) = 0\cdot95.$$

As for means, so also for variances it is often more useful to find a confidence interval than to test a particular value of σ^2 in a Null Hypothesis. Example 11.4.1 shows both calculations on the same data.

11.4.1 Example

Fifteen bean plants are grown in pots in a greenhouse, and their heights are measured after a standard period of growth. In previous experiments, the variance of height has been 27·45 cm^2; for these 15 plants the standard deviation of height is 5·81 cm. (a) Test the N.H. that the variance has not changed; (b) find a 95% confidence interval for σ^2.

(a) A hypothesis test of '$\sigma^2 = 27\cdot45$' against an A.H. 'σ^2 is not equal to 27·45' requires the calculation of $\frac{(n-1)s^2}{\sigma^2}$; this is $\frac{(14\times5\cdot81^2)}{27\cdot45}$ and is equal to 17·216. The values of χ_L^2 and χ_U^2, with 14 d.f., from Table II, are 5·629 and 26·12. The calculated value is in between them, i.e. is within the central 95% of $\chi_{(14)}^2$. The N.H. cannot therefore be rejected on this evidence.

(b) A 95% confidence interval for σ^2, based on this sample of data, is found by calculating the two limits given above: $\frac{(n-1)s^2}{\chi_U^2} = \frac{(14\times5\cdot81^2)}{26\cdot12} = 18\cdot09$, and $\frac{(n-1)s^2}{\chi_L^2} = \frac{(14\times5\cdot81^2)}{5\cdot629} = 83\cdot96$. With probability 0·95, the interval 18·09 to 83·96 therefore contains the true value of σ^2. This interval appears very wide, but we must remember that σ^2 is in units of x^2; the corresponding standard deviations are 4·25 to 9·16.

The value $\sigma^2 = 27\cdot45$ is well within the confidence interval (b), as it must be since the hypothesis test (a) did not give a significant result.

11.5 Comparing two variances: the F distribution

Suppose that two samples, of n_1 and n_2 observations respectively, are drawn from normal distributions, and the sample variances s_1^2, s_2^2 calculated; nothing need be known about the means in order to do this (they are, of course, also estimated from the samples). The variance-ratio, that is the ratio of the two variances estimated from the samples, s_1^2/s_2^2, follows the distribution known as F. This distribution is very similar in shape to that of χ^2, taking positive values only, and being skew with a long tail on the right (as is χ^2 in Fig. 11.1); but since it depends on two variable items, s_1^2 and s_2^2, it will have two parameters, namely the degrees of freedom (n_1-1) and (n_2-1). This causes difficulty in preparing tables, and the course usually adopted (as in Table III) is to present, for only a few chosen levels of probability, the values of F which are exceeded with the given probability.

If we set up the N.H. that the samples came from populations with the same variance (but not necessarily the same mean), the expected value of $F = s_1^2/s_2^2$ is 1. In the most common application of F, the Analysis of Variance (Chapters 14–17), an A.H. is specified which demands a one-tailed significance test; if s_1^2 is not the same as s_2^2 it will have to be greater and cannot be less except through sampling variation. For this reason, F-tables, like χ^2-tables, are always presented in one-tailed form. But in the direct use of F for comparing two sample-variances, a two-tailed test is needed, so that besides the commonly quoted 5%, 1% and 0·1% points, Table III quotes $2\frac{1}{2}$% points.

The ratio

$$s_1^2/s_2^2 \text{ is } F_{[(n_1-1),(n_2-1)]} \text{ and } s_2^2/s_1^2 \text{ is } F_{[(n_2-1),(n_1-1)]}.$$

In testing, it is not necessary for the larger d.f. to be placed first; but what *is* necessary (because of the limitations of the Tables) is to arrange that the numerical value of F shall not be less than 1, or in other words the larger s^2 is divided by the smaller. To see if the result of this is significant at one of the tabulated levels, e.g. 5%, look along the top row of Table III to find the first d.f., then down that column until the row corresponding to the second d.f. is reached; the entry at that point in the table is the F-value, with that pair of d.f., that just achieves 5% significance. For example, $F_{(8,12)} = 2·85$, $F_{(1,24)} = 4·26$, $F_{(9,3)} = 8·81$ and $F_{(15,5)} = 4·62$ at 5%.

11.5.1 Test
Two samples are given, of sizes n_1, n_2 from normal distributions of unknown means. Test the N.H. that the distributions have the same variance. Calculate s_1^2 (with (n_1-1) d.f.) and s_2^2 (with (n_2-1) d.f.). Suppose $s_1^2 > s_2^2$. Set $F_{[(n_1-1),(n_2-1)]} = s_1^2/s_2^2$ and reject the N.H. only if F is significantly large, using a two-tailed test (i.e. upper $2\frac{1}{2}$% etc., not 5% etc., points).

11.5.2 Example
Two samples have been drawn at random from normal distributions; for the first, of $n_1 = 15$ observations, the mean $\bar{x}_1$ is 6·35 and the variance s_1^2 is 17·2074, while for the second sample $n_2 = 20$, $\bar{x}_2 = 4·18$, $s_2^2 = 7·0475$. Test whether the samples could have come from populations with the same variance.

Note first of all that the means are rather different, and could even be significantly so, but this does not affect the test. The ratio $s_1^2/s_2^2 = 17·2074/7·0475 = 2·44$; this is distributed as $F_{(14,19)}$. The N.H. for this problem is that the variances are equal, and it is being tested only against the vague alternative of inequality, no direction for the inequality being specified; this demands a *two*-tailed test of significance and we have already noted that Table III contains those values of F that are used in *one*-tailed tests. All that we ask of our F-ratio here is that it shall not exceed the upper $2\frac{1}{2}$% point (nor be less than the lower $2\frac{1}{2}$% point, but this cannot occur since we have arranged that F shall be greater than 1). The upper $2\frac{1}{2}$% point for F with 14 and 19 d.f. is 2·65; our value is less than this and so not significant, allowing us to accept that the two populations did have the same variance.

(Had we obtained the same numerical value, 2·44, for $F_{(14,19)}$ in an Analysis of Variance where a one-tailed test is needed, it would have to be compared with the 5% point in Table III; this is 2·26, and our value exceeds it, so that a one-tailed test would have rejected the N.H.)

11.5.3 Example. 'Pooling' variance estimates
In Example 8.7.4, we computed a 'pooled' variance estimated from two independent samples of data, prior to carrying out a *t*-test of the difference between their means.

Before doing this, the F-test must be used to check that the two estimates of variance do not differ significantly from each other; i.e. the N.H. is $\sigma_1^2 = \sigma_2^2$.

The estimates were $s_1^2 = \frac{48 \cdot 8800}{6} = 8 \cdot 1467$ on 6 d.f., and $s_2^2 = \frac{39 \cdot 8685}{7} = 5 \cdot 6955$ on 7 d.f. The ratio $s_1^2/s_2^2 = 1 \cdot 43$, and is distributed as $F_{(6,7)}$. The upper 5% point of $F_{(6,7)}$ is 3·87, but since the A.H. does not specify which of these variances should be the larger we should use the $2\frac{1}{2}$% point which is 5·12. Clearly there is no serious evidence against the N.H., and it was perfectly satisfactory to combine all the data and obtain a common s^2 as in Example 8.7.4.

11.6 Exercises on Chapter 11

11.6.1 The height to which seedlings of two varieties of cider apple grow under standard conditions in a greenhouse is assumed to be normally distributed. Ten seedlings from one variety grow respectively 44, 26, 1, 79, 53, 38, 62, 80, 33, 13 cm, and 12 from the other variety grow 33, 47, 55, 39, 24, 61, 38, 12, 26, 64, 52, 51 cm.

(a) Test the hypothesis that the variance of height is the same in the two populations.
(b) Set 95% confidence limits to the estimates of variance for each variety.

11.6.2 Two propagation methods gave the following measurements (mm) of seedling growth after a fixed time:

I:	17·2, 5·1, 12·3, 8·2, 13·5, 13·3, 11·6, 15·2, 10·8, 7·1
II:	19·0, 12·4, 17·5, 12·8, 13·0, 10·6, 4·7, 9·3, 16·8, 10·1

(a) Draw a dot-plot for each method I, II.
(b) Test for each method the hypothesis that the true value of variance in the population is 15.
(c) Set 95% confidence limits to the values of the variances for the two methods.
(d) Comment on the results of this analysis.

11.6.3 For the data on growing plants quoted in Exercise 9.6.8, complete the comparison of the two soils by testing the hypothesis that the variances are the same. Comment on all the results you have found.

12 Tests using the chi-squared distribution

12.1 Using the normal approximation for proportions

There are several tests by which we may study data that are expressed as proportions of observations falling into various classes. They depend on using the normal approximation to the distribution of a proportion, which we have already seen in Chapters 7–9. The goodness-of-fit of a set of data to a hypothesis that they ought to follow, for example, a binomial or a Poisson distribution is assessed by comparing the proportion of 1s, 2s, 3s etc. in the sample with those which ought to arise if the hypothesis were true. In the same way a genetic ratio hypothesis provides a base against which to look at a sample of data.

Sometimes we want to compare sets of classified data from different sources, such as the same genetic ratio hypothesis being examined in groups of animals of related, but not exactly the same, species. Is the ratio the same in species 1 as in species 2? The task of comparing two observed proportions p_1 and p_2 can be set out in the same way, and these two problems are called *contingency table* analysis.

We will consider each of these problems in this chapter; all will use the chi-squared distribution as the basis of the statistical test, but the rules for finding the correct *degrees of freedom* to use must be noted very carefully.

12.2 Goodness-of-fit of data to a ratio hypothesis

Hypotheses in Genetics often lead to ratios of different types to be expected in a population. For example, a characteristic depending on two genes which have dominant or recessive forms will give a 9:3:3:1 ratio for the types AB, Ab, aB, ab. If all four types are distinctive and the double recessive is viable, a sample can be examined and each member of it classified into one of four types. This gives a table of (O_i), observed frequencies. The corresponding (E_i) must add up to the same total as the (O_i) and must be in the ratio 9:3:3:1. In real data, we do not expect the (O_i) to be in *exactly* the right ratio, but we can check whether they are so within acceptable statistical limits.

12.2.1 Test

A sample of n observations is given, each one of which can be put into one of r categories; the frequencies observed in each category are $O_1, O_2, \ldots, O_r$. Calculate how the frequencies would be expected to fall into the categories upon the given hypothesis; call these frequencies $E_1, E_2, \ldots, E_r$. Then

$$\chi^2_{(r-1)} = \sum_{i=1}^{r} \frac{(O_i - E_i)^2}{E_i}.$$

12.2.2 Example

A set of data is expected to show types AB, Ab, aB, ab in the ratio 9:3:3:1. A sample of 556 observations gave totals 315, 101, 108 and 32 respectively in the four groups. Test whether this agrees with the given ratio.

If the given ratio operates, $\frac{9}{16}$ of the total number of observations should fall in the first group, $\frac{3}{16}$ in the second, $\frac{3}{16}$ in the third, $\frac{1}{16}$ in the fourth: these *expected* numbers must add to the same total as the observed frequencies.

The general rule for degrees of freedom in a χ^2 test is that we take the number of items of information, which is the number of pairs (O_i, E_i), and subtract the number of *constraints*, or restrictions, put on the E_i. In this test there is one constraint: the sum of the E_i must be the same as the sum of the O_i, namely the total frequency.

Construct the table

Type	AB	Ab	aB	ab	Total
Observed frequencies	315	101	108	32	556
Expected frequencies	312·75	104·25	104·25	34·75	556

Thus

$$\chi^2_{(3)} = \frac{(315 - 312\cdot75)^2}{312\cdot75} + \frac{(101 - 104\cdot25)^2}{104\cdot25} + \frac{(108 - 104\cdot25)^2}{104\cdot25} + \frac{(32 - 34\cdot75)^2}{34\cdot75}$$

$$= \frac{2\cdot25^2}{312\cdot75} + \frac{3\cdot25^2}{104\cdot25} + \frac{3\cdot75^2}{104\cdot25} + \frac{2\cdot75^2}{34\cdot75} = 0\cdot47.$$

A one-tailed test must be used because very small values of χ^2 would indicate a very close agreement with the null hypothesis, and certainly should not lead us to reject it. The value of 0·47 in this example is very small, and we can see from the table that (O_i) and (E_i) agree well in each type. Only if χ^2 had been significantly large (greater than 7·81 for 5% significance with 3 d.f.) could we have rejected the 9:3:3:1 ratio in the Null Hypothesis.

12.3 Contingency tables

When each member of a population has been examined for two characteristics, and each characteristic classified into a number of categories, we may want to know if the two characteristics are independent. For example, if each of the plants in a long row is examined for leaf colour (from green to yellow) and vigour (good to weak), it may seem likely that plants with yellow leaves are less vigorous. We can set up a N.H. that leaf colour and vigour are *independent* and attempt to discredit this.

12.3.1 Test

Suppose that two characteristics A, B are examined on several individuals. A can take any of the forms $A_1, A_2, A_3, \ldots, A_r$ and B any of $B_1, B_2, \ldots, B_c$. The number of individuals for which A is type A_i, and B is type B_j is denoted by the symbol O_{ij}. A table of these observed numbers is constructed, and the row and column totals a_i ($i = 1$

to r) and b_j ($j=1$ to c) found, as well as the grand total n. This table is called an $r \times c$ contingency table (it has r rows and c columns). If A and B are independent, the number of individuals to be expected in (A_i, B_j) is $a_i b_j / n = E_{ij}$. Test

$$\sum_{\text{all } i} \sum_{\text{all } j} \frac{(O_{ij} - E_{ij})^2}{E_{ij}} \text{ as } \chi^2_{(r-1)(c-1)}.$$

If this is significant, reject the N.H. that A, B are independent.

Char. A	Char. B				Row totals
	B_1	B_2		B_c	
A_1	O_{11}	O_{12}	...	O_{1c}	a_1
A_2	O_{21}	O_{22}	...	O_{2c}	a_2
A_3	O_{31}	O_{32}	...	O_{3c}	a_3
	.	.		.	.
A_r	O_{r1}	O_{r2}	...	O_{rc}	a_r
Column totals	b_1	b_2		b_c	n

The formula for E_{ij} arises in the following way. If A, B are independent, then the ratios of $B_1:B_2:B_3: \ldots :B_c$ should be the same on any A row, and hence the same in the totals $b_1:b_2:b_3: \ldots :b_c$; thus in row A_1, a proportion b_1/n of a_1 should be in the first column, b_2/n of a_1 in the second, and so on.

Another way of thinking of degrees of freedom (an alternative to what we said in Section 12.1 which gives the same answer) is that once a certain number of the (E_i) have been calculated the remainder must follow because of the restrictions or *constraints* imposed.

In a table with r rows and c columns we begin with rc pairs (O, E). In each row *except the last* there are $(c-1)$ observations which have 'freedom', but then the last one is already determined because the row total has to add up. What the contingency table analysis does is to re-allocate the observed total numbers in the rows and in the columns so that the observed totals follow the ratios or proportions specified by the Null Hypothesis: the row and column totals themselves cannot be altered. When we reach the last row all the Es are already determined because the column totals have to add up; so there are $(r-1)$ rows in each of which $(c-1)$ observations need to be calculated. After that the remainder of the entries in the (E_{ij}) table can be written down by making the row and column totals equal the corresponding totals in the (O_{ij}) table.

12.3.2 Example

A sample of 250 seedlings is classified for vigour and leaf colour, with the results below. Test whether these two characteristics are independent.

Observed results:

Leaf colour	Vigour			Total
	Good	Average	Weak	
Green	55	79	4	138
Yellow-green	11	60	15	86
Yellow	1	6	19	26
	67	145	38	250

The expected entries in this table, if colour and vigour are independent, are: Green/ Good $(138 \times 67)/250 = 37 \cdot 0$; Green/Average $(138 \times 145)/250 = 80 \cdot 00$; Green/Weak $(138 \times 38)/250 = 21 \cdot 0$; and so on, giving the Expected table:

	Good	Average	Weak	Total
Green	37·0	80·0	21·0	138
Yellow-green	23·0	49·9	13·1	86
Yellow	7·0	15·1	3·9	26
	67	145	38	250

Now $(r-1)(c-1) = 2 \times 2$, so that χ^2 has 4 d.f. For Green/Good $(\text{Obs} - \text{Exp})^2/$ Exp $= (55 - 37 \cdot 0)^2/37 \cdot 0 = (18 \cdot 0)^2/37 \cdot 0$; for Green/Average $(79 - 80 \cdot 0)^2/80 \cdot 0$, etc. Adding all nine terms from all nine cells in the table, we obtain

$$\chi^2_{(4)} = \frac{(18 \cdot 0)^2}{37 \cdot 0} + \frac{(1 \cdot 0)^2}{80 \cdot 0} + \frac{(17 \cdot 0)^2}{21 \cdot 0} + \frac{(12 \cdot 0)^2}{23 \cdot 0}$$
$$+ \frac{(10 \cdot 1)^2}{49 \cdot 9} + \frac{(1 \cdot 9)^2}{13 \cdot 1} + \frac{(6 \cdot 0)^2}{7 \cdot 0} + \frac{(9 \cdot 1)^2}{15 \cdot 1} + \frac{(15 \cdot 1)^2}{3 \cdot 9} = 100 \cdot 2.$$

This is significant at even more than the 0·1% level, so we cannot accept the hypothesis that the colour and vigour are independent; inspection of the tables reveals that there are far more yellow/weak plants observed than would be expected.

When data are classified, rather than measured, and especially if more than one observer is involved, it is essential to have rigid definitions of the different categories— what is 'average', what is 'yellow-green'. Insufficient control of data collection can render any analysis pointless.

12.4 Proportions

A common type of statement to be found in reports and in advertising runs like this: 'Our new, improved strain of seed gives 90% germination, as compared with 80% which is normal for seed of this plant'.

First we must ask how much data the 80% figure is based on, and whether this can be treated as a 'fixed' value like the ratio in Section 12.2. If so, we have an alternative to the test based on the normal approximation to a binomial variable, and to a proportion, in Section 8.8. We must know how many observations there were made on the new strain, otherwise we cannot complete either that test or the present alternative. The observations on each seed of the new strain are simply classified as 'germinated' or 'not germinated'.

12.4.1 Example
Suppose that 300 of the new seeds were planted and 270 of them germinated. Is this an improvement on the existing 80% rate of germination? We must set up a Null Hypothesis that it is not; then we can calculate some Expected figures against which to

compare our Observed data. The (O_i) are 270 and 30, with corresponding (E_i) of 240 and 60. These are set out in a table:

		Germinated	Not germinated	Total
Standard (old strain) (E_i):		240	60	300
New strain	(O_i):	270	30	300

$$\sum \frac{(O_i - E_i)^2}{E_i} = \frac{(270-240)^2}{240} + \frac{(30-60)^2}{60} = \frac{30^2}{240} + \frac{(-30)^2}{60} = 18 \cdot 75.$$

χ^2 has 1 degree of freedom because there are only two pairs (O_i, E_i) and there is one restriction on the Es—their sum must be 300. This is a very highly significant value and the N.H. must be rejected in favour of an A.H. that the new strain shows a higher germination rate.

But there may not be a large amount of past data collected in similar conditions to those in which the new strain has been grown. Now we have no grounds for setting up a figure such as 80% as a 'standard', but must conduct an experiment to grow a number of seeds of both strains together. The analysis of this must follow that of Section 12.3, where both sets of data have the same status; this is equivalent to comparing two estimated proportions $\hat{p}_1$ and $\hat{p}_2$.

12.4.2 Example

Suppose now that under comparable growing conditions 90 out of 100 seeds of the new strain germinated while 160 out of 200 of the old strain did so. The (O_{ij}) table is:

	Germinated	Not germinated	Total
New strain	90	10	100
Old strain	160	40	200
	250	50	300

Corresponding entries E_{ij} are $\frac{250 \times 100}{300} = 83 \cdot 33$ for 'new, germinated', $\frac{250 \times 200}{300} = 166 \cdot 67$ for 'old, germinated'; and the other two entries are found in the same way. Because this table has two rows and two columns, there will again be 1 d.f. for χ^2.

$$\chi^2_{(1)} = \frac{(90 - 83 \cdot 33)^2}{83 \cdot 33} + \frac{(10 - 16 \cdot 67)^2}{16 \cdot 67} + \frac{(160 - 166 \cdot 67)^2}{166 \cdot 67} + \frac{(40 - 33 \cdot 33)^2}{33 \cdot 33} = 4 \cdot 80.$$

This is greater than the 5% point (which is 3·84) but not so great as the 1% point (6·64). Hence we reject, at the 5% level, the N.H. that the germination rates do not differ significantly, and thus accept that the new strain appears to be better.

Because of the relatively small number of observations, we have a result which, though significant, is not nearly so significant as that in Example 12.4.1 where we assumed that the old strain rate was 'known' to be 80%. We were also able to devote all the experimental resources to examining the new strain, which increased the precision with which its percentage germination was determined.

An alternative way of carrying out this comparison, which is the basis of some computer routines, is set out in Test 12.4.3, and we will illustrate it using the same data as in Example 12.4.2.

12.4.3 Test

In a sample of $(a+b)$ items, a of them possess a certain characteristic and b do not; in a second sample of $(c+d)$ items, c possess the characteristic and d do not. To test the N.H. that the proportions $a/(a+b)$ and $c/(c+d)$ are not significantly different, arrange the data in table:

Observed nos	With characteristic	Without characteristic	Total
Sample 1	a	b	$(a+b)$
Sample 2	c	d	$(c+d)$
	$a+c$	$b+d$	n

$n = a+b+c+d$. Test as $\chi^2_{(1)}$:

$$\frac{n(ad-bc)^2}{(a+b)(a+c)(b+d)(c+d)}$$

using a one-tailed test.

12.4.2 Example (repeated).

$$ad-bc = (90 \times 40) - (10 \times 160) = 3600 - 1600 = 2000.$$

Thus

$$\chi^2_{(1)} = \frac{300 \times 2000 \times 2000}{100 \times 200 \times 250 \times 50} = 4 \cdot 80.$$

Note 1 The result of the test depends not only on the proportions compared but on the number of observations available; this number, n, has to be much larger than the sample sizes required in t-tests, because the information supplied by each observation is less detailed, being only a classification into one of two types, not an exact measurement.

Note 2 If the table contains any entries that are much smaller than 5, *Yates' correction* may improve the quality of the approximation upon which the test is based. This correction reduces the value of $(ad-bc)$ numerically by $\frac{1}{2}n$ before squaring, so that

$$\chi^2_{(1)} = \frac{n(|ad-bc| - \frac{1}{2}n)^2}{(a+b)(c+d)(a+c)(b+d)}.$$

When applying Yates' correction to the first form of the analysis described for Example 12.4.2, each different $(O_{ij} - E_{ij})$ is reduced numerically by $\frac{1}{2}$ before squaring. Clearly this reduces the value of χ^2, and so reduces the danger of obtaining 'false' significance because of the test being approximate.

12.5 Goodness-of-fit of data to a distribution

12.5.1 Example
Data on the numbers of male rats in 100 litters, of five rats each, were:

Number of males, r:	0	1	2	3	4	5	Total
No. of litters(O_i), f:	3	13	30	33	17	4	100

If the conditions stated in Chapter 5 are satisfied, these data should follow a binomial distribution with $n=5$ (the fixed sample, or litter, size). We may *either* assume $p=\frac{1}{2}$, which is not necessarily true, *or* estimate it from the data.

The sample mean $\bar{r} = \frac{(3\times0)+(13\times1)+(30\times2)+(33\times3)+(17\times4)+(4\times5)}{100} = \frac{260}{100} = 2\cdot60$.

Setting $n\,\hat{p}=\bar{r}$, we find $\hat{p}=0\cdot52$.

Frequencies on the binomial hypothesis with $n=5$, $p=0\cdot52$ are

$$100 \times \frac{5!}{r!(5-r)!}(0\cdot52)^r(0\cdot48)^{5-r} \text{ for } r = 0, 1, \ldots, 5.$$

These will form the *expected* frequencies (E_i).

r:	0	1	2	3	4	5	Total
(E_i)	2·55	13·80	29·90	32·40	17·55	3·80	100

Visually from a bar diagram the O_i and E_i would look very similar. To test this, calculate

$$\sum_{\text{all } i} \frac{(O_i - E_i)^2}{E_i} = \frac{(3 - 2\cdot55)^2}{2\cdot55} + \frac{(13 - 13\cdot80)^2}{13\cdot80} + \frac{(30 - 29\cdot90)^2}{29\cdot90} + \frac{(33 - 32\cdot40)^2}{32\cdot40}$$

$$+ \frac{(17 - 17\cdot55)^2}{17\cdot55} + \frac{(4 - 3\cdot80)^2}{3\cdot80} = 0\cdot165.$$

The test compares each observed frequency with its corresponding expected frequency. The expression we have calculated gives a value that follows a χ^2 distribution. There were six pairs (O_i, E_i) and the sum of the Es was constrained to be 100. In this example we were not given any hypothetical value for p, and it had to be estimated from the data before the (E_i) could be calculated. This also constrains the *mean* of the (E_i) to equal that of the (O_i). So d.f. $=6-2=4$, because (E_i) have two constraints on them. (Note that if we had been told $p=\frac{1}{2}$, there would only be one constraint and so 5 d.f.)

The N.H. is 'data are binomial with $p=0\cdot52$', the A.H. is that they are not. The 5% point of $\chi^2_{(4)}$ is 9·49, and so we cannot reject the N.H. This could mean that births were independent as far as sex is concerned, with a slight bias towards each birth being male ($p = 0.52$, not 0.50).

In the case where a binomial model fails to fit, a bar-diagram can help to show what sort of departures from the necessary conditions are occurring.

The χ^2 test is only approximate. If too many (E_i) fall below 5, the approximation is poor. In such cases some groups have to be combined. Some books would recommend combining $r = 0$ with $r = 1$ in Example 12.5.1 and perhaps also $r = 5$ with $r = 4$. However, this would leave very few d.f. for the χ^2 test and make it less sensitive to departures from a binomial pattern. A good compromise is 'not too many expected frequencies should be far below 5'.

12.5.2 Example. Poisson-distributed data

The problem of small expected frequencies will always arise when dealing with data which follow a Poisson model, where there is no theoretical upper limit to the value r may take; the last entry has to be '$r \geqslant$ some value'. Some data on the number of printing errors per page for 100 pages of a magazine were:

Errors per page, r:	0	1	2	3	4	5	6	7	Total
No. of pages (O_i), f:	8	18	30	25	12	4	1	2	100

The expected frequencies, calculated using the mean of the data, which is $\bar{r} = 2.41$, come from the formula $100 \times e^{-2.41} (2.41)^r / r!$ and are

r:	0	1	2	3	4	5	$\geqslant 6$	Total
(E_i):	8.98	21.65	26.08	20.95	12.62	6.08	3.64	100

Strict application of the 'not less than 5' rule would make $r \geqslant 5$ the top group. But it seems unnecessary to lose another d.f. to achieve this. There are now seven pairs (O_i, E_i), the last (O_i) entry being 3. Constraints are 'one for the total' (as always) and one for the estimated mean, leaving $7 - 2 = 5$ d.f.

$$\sum_{\text{all } i} \frac{(O_i - E_i)^2}{E_i} = \frac{(-0.98)^2}{8.98} + \frac{(-3.65)^2}{21.65} + \frac{3.92^2}{26.08} + \frac{4.05^2}{20.95} + \frac{(-0.62)^2}{12.62} +$$

$$\frac{(-2.08)^2}{6.08} + \frac{(-0.64)^2}{3.64} = 2.949.$$

The 5% point of χ^2 is 11.07, so we cannot reject a Null Hypothesis that the data follow a Poisson distribution. This will often be taken to show that conditions of randomness, independence and constant rate (of errors in this example) are satisfied. Strictly speaking the argument should run the other way: if those conditions hold then data will follow a Poisson distribution, but there may be other ways in which the Poisson could arise.

12.6 MINITAB

The only test that can be done directly is that for contingency tables. Others have to be built up using the available mathematical instructions.

CHISQUARE will take data in *r* rows and *c* columns and test for independence of these two classifications. Output is a table showing the O_{ij}, with corresponding E_{ij} directly beneath, the row and column totals, the d.f. for the χ^2 test, and the individual contributions to χ^2 from each (*ij*) combination. This is useful when looking for reasons for significance. It also warns if there are expected frequencies below 5.

12.7 Exercises on Chapter 12

12.7.1   Complete Exercise 6.5.6 by using a suitable test.

12.7.2   Compare the success-rates of the two preparations A and B, described in Exercise 9.6.10.

12.7.3   One page in a table of random numbers contains 800 digits, and on this page the frequencies of the digits 0, 1, 2, . . . , 9 are counted.

Digit	0	1	2	3	4	5	6	7	8	9	Total
Frequency	85	77	83	90	69	79	80	76	84	77	800

Do these results contradict the hypothesis that each digit is equally likely to occur at any entry in the table?

12.7.4   Three strains of corn are thought to be the same genetically. A number of seedlings of each strain are grown, and classified as type *a, b, c, d* in respect of a certain characteristic. The results are:

Strain	Type *a*	*b*	*c*	*d*	Total examined
I	75	15	25	5	120
II	85	37	26	12	160
III	60	28	19	13	120

Test whether the strains are the same in respect of this characteristic. Also test whether the ratio of *a* to (*b* + *c* + *d*) in Strain II could be 9:7.

12.7.5   Presence and absence of two species of plant in 100 sampling quadrats was recorded. Test the hypothesis that the two species occur independently of one another.

		Aster tripolium Present	Absent
Atriplex	Present	29	12
lastata	Absent	34	25

12.7.6   Notification in London of poliomyelitis and polioencephalitis in 1937–46 and in the epidemic of 1947 are summarised below. Examine whether the age distribution of those attacked in the epidemic year showed a significant variation from previous experience.

	Number of notifications	
Age of those attacked	1937–46	1947
0–14 years	467	453
15 and over	131	249
Total	598	702

12.7.7 In a cross involving two Mendelian factors we expect to obtain four classes in the ratio 9:3:3:1 when we interbreed the hybrid (F_1) generation, on the hypothesis that the two factors segregate independently and the four classes of offspring are equally viable. Test whether the following data on Primula agree with this hypothesis.

Flat leaves		Crimped leaves		
Normal Eye	Primrose Queen Eye	Lee's Eye	Primrose Queen Eye	Total
328	122	77	33	560

12.7.8 The following data are extracted from records of patients suffering peptic ulcers or gastric cancer, and also from 'control' patients in hospital for other reasons. Test whether blood types are equally affected by these diseases.

Blood type	Peptic ulcer	Gastric cancer	Controls
O	55	21	162
A	37	24	154
B	8	5	34

12.7.9 A true-breeding stock of ebony-bodied flies (*Drosophila*) was crossed to a normal stock. The progeny were crossed with one another to produce a second generation. Ebony-bodied colour is due to a recessive gene (e), so that crossing e/e and +/+ parents (+ means normal) will give e/+ in the first generation. The second generation gives +/+, +/e, e/e in the ratio 1:2:1, only the last group of which appear ebony. In the actual experiment, 27 out of 90 were ebony. Test whether these data fit the ratio.

12.7.10 The numbers of snails *spirorbis corallinae* on quadrats 10×10 mm on the seaweed *laminaria digitata* were counted, as follows:

Number of snails	0	1	2	3	4	$\geqslant 5$
Number of quadrats	27	35	24	18	8	8

Using all the data (before combining the class $\geqslant 5$), the mean was 1·8 per quadrat. Test whether these observations can be explained by a Poisson distribution.

(In a Poisson distribution with mean 1·8, the probabilities of values of r are:

r:	0	1	2	3	4	$\geqslant 5$
$P(r)$:	0·1653	0·2975	0.2678	0·1607	0·0723	0·0364).

12.7.11   In a test of electronic components for which the failure rate is thought to be constant the numbers of failures in a standard test are:

Number of failures, r:	0	1	2	3	4	5	6	Total
Frequency, f:	26	30	26	18	9	5	6	120

By suitable diagrams and tests, examine the hypothesis that failures occur independently and at random.

12.7.12   Apples are packed in containers for marketing, and each container holds three apples. Although the fruit are checked visually for damage before packing, there is a probability of $\frac{1}{6}$ that any apple will have developed a soft spot by the time it is sold.

An inspector checks 216 containers on the shelves of supermarkets and finds the following information.

Number of soft fruit	0	1	2	3
Number of containers	110	85	20	1

Should he accept the claim that the probability of finding a soft spot on an apple is no more than $\frac{1}{6}$?

13 Correlation between two variables

13.1 Pairs of measurements

So far we have been concerned entirely with populations wherein one measurement only has been taken on each member; we have called it x_i or r_i. Sometimes measurements might have been taken before and after treatment, but they would be of the same characteristic (e.g. weight of an animal) and most probably the really interesting variate would be the difference between the two. Essentially, one single characteristic of the population has been under study.

There are occasions when two measurements, x_i and y_i, are taken simultaneously on each member of a population, for the purpose of seeing whether they are related. The reason for this may be pure scientific interest, such as measuring the height and weight of each member of an animal population to see if any general laws or patterns of relationship between height and weight can be set up; or it may be the search for measurements y_i, which are easier to take than the x_i that are really of interest, and are at the same time sufficiently closely related to the x_i to give the same information. Thus it is common to measure the amount of a chemical present in a solution, x_i, by the optical density of the solution, y_i, since the latter can be determined at once on a standard laboratory instrument once this has been suitably calibrated (Example 14.1.1): we do not wish to know y_i, but it is a convenient short cut to x_i, in which we are really interested.

Correlation coefficients used to be calculated almost whenever an opportunity presented itself, but in recent work they have taken a much less prominent place as more detailed methods of examining relationship have been developed: some of these will be studied in later Chapters.

13.2 Scatter diagrams

The first step in studying possible relationships between two measurements x and y is to draw a **scatter diagram** (or *scattergram*). Figure 13.1 shows various types of relationship, and these can be related to the results of calculations which will be described in Section 13.3. Each cross represents a unit (animal, plant etc.) on which x and y have been measured; the coordinates of the cross representing the ith unit on the graph are (x_i, y_i). An additional useful piece of information which may be shown on the graph is the position of each of the means, $\bar{x}$ and $\bar{y}$, of the data set. Figure 13.2 shows this; in that particular diagram we see that usually the units with large x-values also have large y-values, and smaller xs go with smaller ys. Therefore when $(x_i - \bar{x})$ is positive, i.e. x_i is above the mean, then $(y_i - \bar{y})$ will also be positive; likewise the negative values of $(x_i - \bar{x})$, when x_i is less than the mean, go with negative $(y_i - \bar{y})$. Not every unit shows this, but the great majority do. It is this visual impression which can easily be made into a numerical calculation.

13.3 The correlation coefficient

We first need to define a new expression, the *covariance* of x and y. It depends on the *product* of the two deviations of x and y from the general mean, namely $(x_i - \bar{x})\,(y_i - \bar{y})$.

The formal definition is $cov(x,y) =$

$$\frac{1}{(n-1)} \sum_{i=1}^{n} (x_i - \bar{x})(y_i - \bar{y}),$$

and another form of calculation is

$$\frac{1}{(n-1)} \left[\sum_{i=1}^{n} x_i\, y_i - \frac{G_x\, G_y}{n} \right].$$

The symbols G_x and G_y stand for the two grand totals, $\sum x_i$ and $\sum y_i$, and the expression $\sum_{i=1}^{n} x_i\, y_i$ simply means 'for each member of the chosen sample, multiply the measurement of x by the measurement of y on that member, and add these products xy for all members of the sample'.

We note that the definition is very similar to that of a variance, except that $(x_i - x)^2$ is replaced by $(x_i - \bar{x})\,(y_i - \bar{y})$; so there is a similar form of calculation, with x_i^2 replaced by $x_i y_i$ and $(\sum x_i)^2$ replaced by $G_x G_y$. Clearly when the relation is like that shown in Fig. 13.2, the products $(x_i - \bar{x})\,(y_i - \bar{y})$ will nearly all be positive: the values of $(x_i - \bar{x})$ and $(y_i - \bar{y})$ for the same unit will *either* both be positive *or* both be negative. The few cases where one is positive and the other is negative will give negative products, but these will greatly be outnumbered by the positive ones and the sum of these products, which is the covariance, will be positive and quite large. However, the actual value of the covariance depends on what units each measurement x, y is expressed in: by using cm instead of m for lengths, or g instead of kg for masses, the numerical value of the covariance could be altered greatly.

In order to make the measurement of relationship on a standardised scale, we introduce the *correlation coefficient*,

$$r = \frac{cov(x,\ y)}{s_x s_y},$$

dividing the covariance by the two standard deviations, for x and for y. These standard deviations are always positive but the covariance need not be. A relationship like Fig. 13.1(d) will have positive $(x_i - \bar{x})$ going with negative $(y_i - \bar{y})$, or negative $(x_i - \bar{x})$ with positive $(y_i - \bar{y})$, on most units; this will give a negative covariance. By standardising, the values of r are restricted to lie in the range -1 to $+1$.

The expression for r in terms of the observations can be written

$$r = \frac{\sum (x_i - \bar{x})(y_i - \bar{y})}{\sqrt{\sum (x_i - \bar{x})^2 \cdot \sum (y_i - \bar{y})^2}},$$

and as in the case of variances there is an easier form of calculation:

$$r = \frac{n \sum_{i=1}^{n} x_i y_i - G_x G_y}{\sqrt{(n \sum_{i=1}^{n} x_i^2 - G_x^2)(n \sum_{i=1}^{n} y_i^2 - G_y^2)}}.$$

Figure 13.1 illustrates the way in which values of *r* show the relation between *x* and *y*. In Fig. 13.1(a) *r* will be positive and fairly large, perhaps about $+0.75$; in (b) $r = +1$; in (c) $r = -1$; in (d) *r* is negative and quite large, say about -0.8; in (e) and (f) *r* is approximately 0. A correlation coefficient measures how close the relation between *x* and *y* is to *linearity*; if this relation is perfect as in (b) and (c) *r* takes its maximum possible numerical value: the actual slope of the line can be found (as in Chapter 14) when we know in what units *x*, *y* are measured.

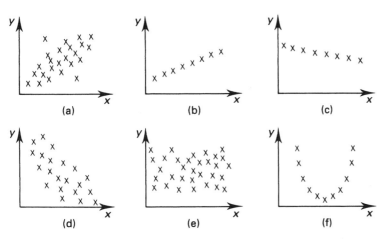

Fig. 13.1 Values of the correlation coefficient, *r*, between *x* and *y*: (a) *r* positive and fairly large; (b) $r = +1$; (c) $r = -1$; (d) *r* negative and fairly large; (e) *r* approximately 0; (f) *r* approximately 0.

In (a) and (d) the points are scattered about a reasonably well-discerned line, while in (e) it is clear that *x* may take any value without restricting the value taken by *y*. But (f) should be a sharp warning against the indiscriminate use of correlation: *r* is virtually zero in cases like this because the values of one variate, in this example *y*, are symmetrically distributed about the mean of the other, *x*. In other words, there are y_i-values of identical size at each of the two points $+(x_i - \bar{x})$ and $-(x_i - \bar{x})$. This symmetrical distribution may be, and often is, a perfectly good, valid and interesting relation between *x* and *y*.

The steps in using the correlation coefficient should thus be: (1) plot the data on a graph (the *scatter diagram*); (2) see if there is any apparently curved relation between the variates; (3) if there is *not* a curve, and particularly when the scatter diagram suggests a *linear* relation, then calculate *r*. The value of *r* in these circumstances will give a valid measure of the size of linear relation.

The significance of the calculated value of *r* may be tested using Table IV, in which

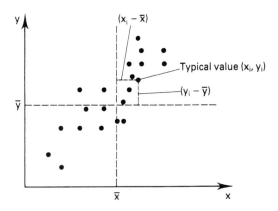

Fig. 13.2   Scatter diagram showing values (x_i, y_i) and the means of $\bar{x}, \bar{y}$ of the complete data set.

the degrees of freedom must be taken as $n-2$ when n pairs of records (x_i, y_i) are available; r must be at least as large numerically (positive or negative) as the value in the Table before it can be called significant. The Null Hypothesis here is that the correlation coefficient in the population is 0; that is, we test whether the calculated value of r (either $+$ or $-$, the sign being left out of the Table) is significantly different from zero. If there were reasons to suggest that the population value was something other than 0, i.e. we were testing a calculated value against a hypothetical value of $\frac{1}{2}$, or $\frac{3}{4}$, or $\frac{1}{3}$, etc., Fisher's z-test would be required; see Clarke and Cooke[3] and Fisher[5] for this, and for testing the difference between two correlation coefficients, calculated from two different samples.

13.4 Interpretation

Various pitfalls exist in interpreting the significance of correlation coefficients, even assuming it was justifiable to calculate them after looking at the data. These occur when the existence of a correlation is then logically equated to a *causal* relation: x causes y or x is due to y. If we are merely looking at x and y to see if they seem to be showing the same trend, so that we then choose to measure only one of them (the more convenient) in order to demonstrate the trend, the logic of cause-and-effect is not so important. But theories have frequently been built on correlations between x and y when these really represented two measurements each quite closely related to some *third* one, z, and not to one another at all directly. The most picturesque examples exist in the social sciences, when z represents time. For example, if x_i denotes the number of television receiving licences taken out in the ith year since 1945, and y_i the number of convicted juvenile delinquents in the same year, this set of measurements showed a considerable, positive correlation over the 20 years 1945–64. Some people would argue that this meant that television had a bad influence on the young; but in fact a similar result could be found by comparing $x_i =$ number of radio licences and $y_i =$ number of people certified insane, over a 15-year pre-war period, and an argument in this case might be harder to sustain! It may well be that all these things are increasing with time (and population) for, perhaps, not entirely unconnected reasons but certainly not as mutual cause and effect.

13.4.1 Example

A group of $n = 20$ strawberry plants was grown in pots in a greenhouse, and measurements were taken on y, the crop yield, and x, the corresponding level of nitrogen present in the leaf at the time of picking:

x:	2·50	2·55	2·54	2·56	2·68	2·55	2·62	2·57	2·63	2·59
y:	247	245	266	277	284	251	275	272	241	265
x:	2·69	2·61	2·67	2·57	2·53	2·70	2·51	2·58	2·53	2·61
y:	281	292	285	274	282	295	249	246	261	260

(x is measured in parts per million by weight of dry leaf matter, and y in g).

$$n = 20, \qquad G_x = 51 \cdot 79, \qquad G_y = 5348,$$

$$\sum_{i=1}^{20} x_i^2 = 134 \cdot 1793, \qquad \sum_{i=1}^{20} y_i^2 = 1\,435\,404, \qquad \sum_{i=1}^{20} x_i y_i = +13\,859 \cdot 60$$

Therefore

$$r = \frac{20 \times 13\,859 \cdot 60 - 51 \cdot 79 \times 5348}{\sqrt{(20 \times 134 \cdot 1793 - (51 \cdot 79)^2)(20 \times 1\,435\,404 - (5348)^2)}} = +0 \cdot 5698.$$

This has 18 d.f., and is significant at the 1% level (being greater than the value of 0·5614 in Table IV), so that we must reject the Null Hypothesis that the population correlation coefficient $= 0$, and accept that there does seem to be a relation between x and y.

This leads to an illustration of another error in interpreting correlation coefficients. Let us suppose that a further set of 50 plants is now examined, but that the fertiliser applied to them contains nitrogen at a level so high that all are likely to be giving their maximum yield. In this case, any variations in x would not be accompanied by equivalent, predictable variations in y, so that no significant correlation would exist. What result would be obtained by working out a correlation coefficient between x and y for the combination of both samples of plants, that is of the whole 70? It would be dominated by the pattern of behaviour in the larger group, the one of 50 plants; these would swamp the smaller group of 20, and the net result would be that apparently x and y are not significantly correlated over the whole set of 70 plants. But this is fallacious because the 70 are not a homogeneous set: they can be broken down into groups (in this case into just two groups) wherein the growth and cropping features are not the same. Had we plotted all 70 points on a scatter diagram, it would have been clear that the same linear relation did not exist throughout. Note also that x and y must represent *different* measurements; e.g. if x is total crop on a plant and y is Grade I crop, x and y are bound to be correlated for mathematical reasons.

If there is no other way of establishing a theory than by using correlation, great care is needed to eliminate all possibility of indirect relationship, that is via a third variable z: the controversy about cigarette smoking and lung cancer shows how much play can be made on this theme. In this particular instance, whatever *third variables* have been suggested, the relation has seemed to remain when they have been eliminated.

There is no straightforward method of setting limits to calculated values of r, particularly for small sample sizes, n, because the distribution of r is not symmetrical enough to use standard-error methods.

The value of r, as a measure of relation between x and y, is a reliable indicator only when x and y are normally distributed, so that it is wise to check that this condition is, at least approximately, satisfied before making the calculation. A wild, 'outlying' observation can completely dominate the calculation.

13.5 Rank correlation

We may not always be able to measure continuous variates x and y on the n members of a population; but instead it may be quite possible to measure an x and a y, each of which can be placed in order for the n members. For example, the judge at the village Flower Show may be faced with 15 pots of strawberry jam that have to be ranked in order 1, 2, 3 and the rest. Important characteristics will include colour and texture (stiff, about right, or runny) and of course flavour: let us concentrate on the first two. Neither of these characteristics can be measured on a continuous scale, but it is very likely that the 15 pots can each be given a position in the rank order 1, 2, . . . for colour (1 is the best, down to 15 which is the worst) and a rank for texture also (1 is the nearest to good texture, down to 15 which is the worst—either extremely stiff or extremely fluid depending on what the contestants have offered!). How well do the rankings for colour agree with those for texture? Is the best for colour the best for texture, or at least are the best four or five for colour also the best four or five for texture; or does there seem to be no relation at all between the two rank orders?

We label the rank order for colour x and that for texture y, and we will assume that the judge has ranked all 15 pots of jam in the following orders.

Pot	a	b	c	d	e	f	g	h	j	k	l	m	n	p	q
Rank for colour, x	10	14	6	1	7	4	9	13	2	12	5	11	15	3	8
Rank for texture, y	4	11	5	6	12	1	14	10	7	13	2	15	9	3	8

In spite of x and y not being continuous measurements, we will calculate the correlation coefficient between them just as though they were. This process is due to Spearman, and the *rank correlation* coefficient so obtained is called Spearman's coefficient, usually denoted by r_s. Of course it was necessary to study the distribution of r_s, which Spearman did, obtaining the results which we shall use later in testing its value.

The easiest way to calculate r_s is first to add a line to the table of records above, giving d_i, the difference in ranks $(x_i - y_i)$ for each member of the sample; these are:

Pot	a	b	c	d	e	f	g	h	j	k	l	m	n	p	q
d_i	6	3	1	-5	-5	3	-5	3	-5	-1	3	-4	6	0	0

There is a useful check on the correctness of our calculation: the sum of the d_i should be 0.

It can be shown, after some algebra (see, for example, Clarke and Cooke,[3] Section 20.9), that Spearman's r_s is now given by

$$r_s = 1 - \frac{6 \sum_{i=1}^{n} d_i^2}{n(n^2 - 1)}$$

in which n is the number of values of d_i available, that is the number of items being ranked. In our sets of records, $n=15$ and so $n(n^2-1)=15\times224=3360$. Also $\sum d_i^2 = 6^2 + 3^2 + 1^2 + \cdots + 6^2 + 0^2 + 0^2 = 226$, so that

$$r_s = 1 - \frac{6\times226}{3360} = 1 - 0\cdot4036 = +0\cdot5964.$$

We shall now wish to test the significance of the value which we have calculated. Being a correlation coefficient, it has to lie between -1 and $+1$; the value $+1$ will occur when the two rankings x and y agree exactly, and -1 when they disagree completely (the best for colour is worst for texture, the second best for colour is the second worst for texture, and so on right through the list). The Null Hypothesis usually needed is that there is *no relation* between the two rankings x, y; in other words if a sample of jam has been ranked high in the list for colour this tells us nothing at all about what to expect for texture—it may be good, bad or indifferent. As with an ordinary correlation coefficient this implies $r_s=0$.

For values of n up to 10, a special table is needed (Table V) which shows what value of r_s (positive or negative) is necessary if it is to be significantly different from 0. For larger n, either we may use the table for the ordinary correlation coefficient (Table IV), or to obtain a slightly better approximation we may use the fact that, for $n>10$, the expression

$$r_s\sqrt{\frac{n-2}{1-r_s^2}}$$

has approximately the t distribution with $(n-2)$ d.f. In our example, therefore, $0\cdot5964\sqrt{13/(1-0\cdot3557)}$ may be tested as $t_{(13)}$. This is $0\cdot5964\times4\cdot4918=2\cdot68$, which is significant at the 5% level, leading us to reject the N.H. of no relation between the two rankings. Although some pots of jam have been given very different rankings for the two characteristics, most are either reasonably high for both or quite low for both.

13.6 MINITAB

The first step in a correlation analysis is to plot a scatter diagram. If data are arranged in three columns, the first being the pair number, the next x and the last y, PLOT will do this. CORRELATION then gives the value of r, for the two columns as named, which could be x, y or their actual names. This command can be extended to give all possible correlation coefficients for three or more columns if several variables have been measured on the same units. Rank correlation is not catered for.

13.7 Exercises on Chapter 13

13.7.1 The blood-clotting times of a number of people were measured before and after taking a certain drink; the times before, x sec, and after, y sec, are recorded below. Calculate the correlation coefficient between x and y.

Subject	A	B	C	D	E	F	G	H	J	K	L	M	N	P	Q
x	175	142	124	168	117	134	167	147	126	104	136	129	178	146	149
y	82	90	126	128	127	54	117	100	91	89	61	134	78	106	99

13.7.2   Measurements of the heights (inches) of brother and sister were made in each of 15 two-child families, with the following results. Calculate the correlation between the two heights.

Family	1	2	3	4	5	6	7	8	9	10	11	12	13	14	15
Brother x	73	70	74	68	70	67	69	71	70	68	69	68	71	73	69
Sister y	69	67	63	66	67	64	66	68	67	65	65	64	66	67	67

13.7.3   'The leaf area of the tomato plants used in the experiment was positively correlated with the height to which the plants grew.' What use could be made of the information in this extract from a report?

13.7.4   A group of scientists collected a number of sea urchins from different locations. At each collecting point they noted the mean summer temperature, $x(°C)$ and the average magnesium content of the specimens, $y(\% MgCO_3)$.

x:	18·1	23·0	17·5	20·2	14·7	13·8	15·1	13·8	24·2
y;	8·8	9·5	8·9	9·1	8·6	8·3	8·5	8·2	9·5

Plot these data on a scatter diagram and examine whether there is evidence of a relation between x and y.

13.7.5   Comment on the following statements.

(a) 'The correlation coefficient between x, the mean annual rainfall at an observation station, and d, the station's distance inland, was -0.79 for 38 stations in Sierra Leone. This shows conclusively that rainfall is confined mainly to the coastal strip.'

(b) 'The correlation coefficient between the age, x, of a sow and the number of piglets, y, produced per litter was calculated for all the sows of various breeds on the farm. It was not significantly different from 0, so that (contrary to suspicion) there is no tendency for litter size to decrease with sow's age'.

13.7.6   The percentage of a certain chemical is estimated, in each of 12 samples A–M, by two methods. Method I is much quicker than Method II, but the chemist doing the estimation suspects that it may give less reliable results: in particular he thinks that the figure 89 for E is wrong, and should have been 59. Calculate a *rank* correlation coefficient between the results on the two methods. What advantage might this have over an ordinary correlation coefficient?

Sample	A	B	C	D	E	F	G	H	J	K	L	M
Method I	46	48	55	47	89	45	44	42	50	56	41	49
Method II	42	44	38	36	58	39	37	45	40	47	43	41

13.7.7   Two organisms, X and Y, are each grown in eight different environmental conditions I–VIII. Their sizes after a fixed time are (volumes in cm³):

Environment	I	II	III	IV	V	VI	VII	VIII
Organism X	22	16	38	187	24	68	31	478
Organism Y	44	48	32	155	42	93	35	336

Calculate, and test for significance, a rank correlation coefficient between the sizes of X and Y in the different environments.

The ordinary correlation coefficient is $+0.996$. Comment on the difference between this and the rank coefficient.

13.7.8 Two measurements x, y are made on each of 40 people; x is scored 0 or 1 for each person and y is scored -1 or $+1$. A correlation coefficient is calculated between x and y and is found to be significantly different from 0. What can be inferred from this?

14 Fitting straight lines—linear regression

14.1 Linear relationships

When two variates x, y are for theoretical reasons likely to be related, or when the measurements have been shown on a graph and look like Fig. 13.1(a), (b), (c) or (d), it is natural to try to describe the linear, or approximately linear, relation between them by an equation, the **linear regression** equation. This equation expresses one variate, the *dependent* variate y, in terms of the other, the *independent* variate x. We assume for the moment that it is clear which of the variates is the dependent one, as in Example 14.1.1 where the optical density, y, of a solution depends on the concentration, x, of the chemical present in it, so that we can properly write $y = a + bx$ to show how y alters when x is altered. Figure 14.1 indicates what a and b are: a is the *intercept*, or that value taken by y when x is 0, while b is the *slope* or gradient of the line, the average increase in y for unit increase in x.

We can choose, and fix, the values of x to use in an experiment designed to study this relation, but we have no control over the resulting y. The idea, and the logic, behind calculating a regression line is different from that for correlation: in regression, one measurement y is a variable and depends on the other measurement x which can be fixed, while in correlation neither x or y can be fixed—both are variables.

Example 14.1.1
The optical density, y, of a solution measured at eight concentrations, x, of a chemical was as follows (illustrated in Fig. 14.1):

Meter reading y_i:	4	9	18	20	35	41	47	60	Total $\sum y_i = 234$
Conc. (μg/ml) x_i:	1	2	4	5	8	10	12	15	Total $\sum x_i = 57$

We shall find the line which fits the observed points best, in the sense that for each given value of x_i the value of y_i on the line (the *predicted* value of y_i) is as near as possible to the observed value of y_i. Note that because we are treating x as fixed, no question of *fitting* for x arises.

Consider the sum of the squares of all the vertical deviations like QP on Fig. 14.1 (Q is the observed value of y at $x = 8$, P is the y-value on the fitted line), i.e. consider $S_D = \sum_{i=1}^{n} (y_{i.\text{Obs}} - y_{i.\text{Pred}})^2$ in an obvious notation. The *least-squares* method of estimating a and b involves putting $y_{i.\text{Pred}} = a + bx_i$ and then looking for those values of a, b which will make S_D as small as possible. In other words, the required estimates $\hat{a}$

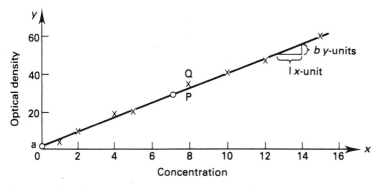

Fig. 14.1 Relation between optical density, y, of a solution and concentration, x, of chemical present in it (Example 14.1.1)

and $\hat{b}$, of a and b, will be those which minimize $S_D = \sum_{i=1}^{n} (y_{i.\text{Obs}} - a - bx_i)^2$. When the necessary mathematics has been done, we find that

$$\hat{b} = \frac{\sum_{i=1}^{n}(x_i - \bar{x})(y_i - \bar{y})}{\sum_{i=1}^{n}(x_i - \bar{x})^2}, \text{ and } \hat{a} = \bar{y} - \hat{b}\bar{x}$$

so that $y = (\bar{y} - \hat{b}\bar{x}) + \hat{b}x$, or more neatly $y - \bar{y} = \hat{b}(x - \bar{x})$. To calculate b, the same sort of algebra is used as for the correlation coefficient, leading to

$$\hat{b} = \frac{[n\sum_{i=1}^{n}x_iy_i - G_xG_y]}{[n\sum_{i=1}^{n}x_i^2 - G_x^2]}.$$

In Example 14.1.1, $n=8$, $G_x=57$, $G_y=234$. Therefore $\bar{x} = 7\cdot125$, $\bar{y} = 29\cdot25$.

$$\sum_{i=1}^{8}x_i^2 = 1^2 + 2^2 + 4^2 + \cdots + 15^2 = 579$$

$$\sum_{i=1}^{8}x_iy_i = (1 \times 4) + (2 \times 9) + \cdots + (15 \times 60) = 2348$$

$$\therefore \hat{b} = \frac{8 \times 2348 - 57 \times 234}{8 \times 579 - 57^2} = \frac{5446}{1383} = 3\cdot94.$$

$\hat{a} = 29\cdot25 - 3\cdot94 \times 7\cdot125 = 1\cdot19$. Therefore the slope of the line is $3\cdot94$, y increasing by this amount for each unit increase in concentration x; and when $x=0$, y takes the value $1\cdot19$, which in this example could represent some sort of zero- or calibration-error of the instrument. To draw the line on the graph, mark the points $(x = 0, y = a)$ and $(x = \bar{x}, y = \bar{y})$, and join them by the regression line: in this case, the two points are those marked by circles in Fig. 14.1. A regression line always passes through the point $(\bar{x}, \bar{y})$. Of course, b, like r, can be negative, if the relation between x and y is as in Fig. 13.1(d). The equation is presented as $y - 29\cdot25 = 3\cdot94(x - 7\cdot125)$, or as $y = 1\cdot19 + 3\cdot94x$. The least-squares technique for finding $\hat{a}$, $\hat{b}$ requires no assumptions about the pattern of variation either in y or in $(y_{\text{Obs}} - y_{\text{Pred}})$; but clearly we need to measure the goodness-of-fit of the fitted line to the values which were actually observed, and some distributional assumptions are required to do this. So far we have laid down only one condition: *that x should be fixed, capable of observation without error* (1). We now add a second, *that the errors in measurement of y shall be independent of one another and of x, and shall be normally distributed with constant variance, σ^2 (2).*

14.2 Goodness-of-fit and Analysis of Variance

The method of testing goodness-of-fit of a linear regression is based on the *Analysis of Variance* which we shall meet in another connection in Chapter 15. The name arises because we use an expression that is very like a variance (*S* below), and analyse it into its constituent parts, each one of which represents some meaningful portion of the whole. Define the *Total Sum of Squares*, $S = \sum_{i=1}^{n}(y_i - \bar{y})^2$; this is a measure of the total variation among all the observed y-values, and if it were divided by $(n-1)$ would be like a variance: it is not quite true to say it would actually be a variance, for we shall see that one part of it can be used to estimate the variance of y while the other part tells us something else.

So far we have thought of the fitted line as $y = a + bx$; it is now clear that each original observed value $y_{i.\mathrm{Obs}}$ is not *exactly* equal to $a + bx_i$, but to this plus a small term representing the deviation of the observed point from the fitted line (as in Fig. 14.1, where QP is a typical one of these deviations). Hence we write $y_i = a + bx_i + e_i$, where y_i and x_i stand for a pair of observations giving a point on the graph, and e_i is the deviation term. The assumption (2) thus requires that the e_i shall be mutually independent, not affected by x and shall all be taken from a $N(0, \sigma^2)$ distribution.

The Analysis of Variance is laid out as in Table 14.1: the first column lists the components into which the total sum of squares is split. These are: one part measuring how much of the variation among the y_i can be explained by the fitted regression line (i.e. the extent to which this variation was merely a linear relation with x), and a second part measuring the size of deviations from this line. It is this second part that is really of interest, for if it is small the line will be a good fit. However, the calculation is most easily made by working out *S* and the sum-of-squares for regression, S_R; and then the difference between these, $S_D(=S-S_R)$, is the required measure of deviation. (It is, in fact, equal to $\sum_{i=1}^{n}(y_{i.\mathrm{Obs}} - y_{i.\mathrm{Pred}})^2$, or $\sum_{i=1}^{n} e_i^2$, though it is not so easily calculated in this way.) *S* has already been defined: it is calculated as $S = (1/n)[n\sum_{i=1}^{n} y_i^2 - G_y^2]$. S_R is found to be $\{\sum_{i=1}^{n}(x_i - \bar{x})(y_i - \bar{y})\}^2 / \{\sum_{i=1}^{n}(x_i - \bar{x})^2\}$, and is calculated as

$$S_R = \frac{[(1/n)(n\sum_{i=1}^{n} x_i y_i - G_x G_y)]^2}{(1/n)(n\sum_{i=1}^{n} x_i^2 - G_x^2)}.$$

For Example 14.1.1, the results are:

$$S = \frac{1}{8}(8 \times 9536 - 234^2) = 21\,532/8 = 2691\cdot5$$

$$S_R = \frac{[\frac{1}{8}(8 \times 2348 - 57 \times 234)]^2}{\frac{1}{8}(8 \times 579 - 57^2)} = \frac{680\cdot75^2}{172\cdot875} = 2680\cdot6685.$$

Then $S_D = S - S_R = 10\cdot8315$.

In Chapter 11, we found that the sum of squares of n unit normal deviates was distributed as χ^2, and that its degrees of freedom were $(n-1)$; further, that if the normal deviates have variance σ^2 rather than unity, the numerical value of this χ^2

divided by its degrees of freedom, $(n-1)$, provides an estimate of σ^2. In the present case, the sum of squares for deviations from regression, S_D, is equal to $\sum_{i=1}^{n} e_i^2$; we have assumed that e_i is $N(0, \sigma^2)$, and therefore S_D is distributed as χ^2. A suitable Null Hypothesis for this problem states that the regression coefficient b is zero. If that is so, S and S_R are also distributed as χ^2; S, being based on n observations, is $\chi^2_{(n-1)}$, but it is not at once clear what will be the d.f. for S_R and S_D. A theorem in the mathematical development of the method of least-squares may be applied to discover that S_D is $\chi^2_{(n-2)}$, and a further theorem, this time about the sums of χ^2 variates, tells us that S_R is therefore $\chi^2_{(1)}$. A good working rule, which applies to more complicated regression equations also, is that S_R has d.f. equal to one less than the number of constants needing to be estimated: in this Example we have estimated a and b. The d.f. of all the components in an Analysis of Variance table (here S_R and S_D) must add to the d.f. of S; hence S_D must have $(n-2)$ d.f.

The second column of Table 14.1 contains the d.f., and the third column the sums of squares S, S_R and S_D. On the N.H., each of the χ^2-variates just mentioned will, when divided by its d.f., give an estimate of σ^2: so the fourth column contains the **mean squares** $S_R/1$, $S_D/(n-2)$ (mean square being the general name for a sum of squares divided by its d.f.). The ratio of two such estimates of σ^2 is distributed as F; specifically, $(S_R/1)/(S_D/(n-2))$ is $F_{(1, n-2)}$. If this F is significantly large, we reject the N.H., and assume that b is not zero. It can be shown that if b is *not* zero, the value of the regression mean square must be greater than σ^2, so the correct F-test is the one-tailed form, as tabulated in Table III. Whether or not b is zero, the deviations-mean-square $S_D/(n-2)$ provides a valid estimate of σ^2, the variance of e_i.

Table 14.1 Algebraic form of Analysis of Variance for testing regression goodness-of-fit

Source of variation	D.F.	S.S.	M.S.	Test
Regression	1	S_R	$M_R = S_R/1$	$M_R/M_D = F_{(1, n-2)}$
Deviations from line	$n-2$	$S_D = S - S_R$	$M_D = S_D/n - 2$	
Total	$n-1$	S		

Table 14.2 shows this method of analysis applied to the data of Example 14.1.1. The ratio of the two mean squares is very large: reference to the F-table with 1 and 6 d.f. shows the actual value to be very much greater than that needed for significance at 0.1%, and so we must reject the N.H. We may claim that b is significantly different from 0.

Table 14.2 Analysis of Variance for Example 14.1.1

Source of variation	D.F.	S.S.	M.S.	
Regression	1	2680·6685	2680·6685	$F_{(1, 6)} = 1484·9$***
Deviations	6	10·8315	1·8053	
Total	7	2691·5000		

In Table 14.2 and later, we use the shorthand that significance at the 5%, 1% and 0·1% levels is indicated by one, two and three asterisks respectively, and n.s. denotes not significant.

14.3 Confidence interval for *b*

The most interesting feature about a regression calculation will usually be the numerical value of the slope *b*, and so it is useful to be able to set limits to *b*. The necessary estimate of σ^2, as we have just seen, is obtained from the deviations-mean-square. Further, the distribution of $\hat{b}$ can be shown to be normal, and so confidence limits can be calculated in the manner described in Chapter 9 for normal variates. The variance of $\hat{b}$ is estimated by $\hat{\sigma}^2/[\sum_{i=1}^{n}(x_i - \bar{x})^2]$, where we use the symbol $\hat{\sigma}^2$ to denote that estimate of σ^2 provided by the deviations-mean-square. This estimate $\hat{\sigma}^2$ has $(n-2)$ d.f., and the rule *estimate plus or minus* **t** *times its standard error* may be applied to give the 95% limits for *b* as

$$\hat{b} - t_{(n-2, 0\cdot05)} \sqrt{\frac{\hat{\sigma}^2}{\sum_{i=1}^{n}(x_i - \bar{x})^2}} \leqslant b \leqslant \hat{b} + t_{(n-2, 0\cdot05)} \sqrt{\frac{\hat{\sigma}^2}{\sum_{i=1}^{n}(x_i - \bar{x})^2}}$$

In Example 14.1.1, $\hat{b} = 3\cdot94$, $\hat{\sigma}^2 = 1\cdot8053$, and $\sum_{i=1}^{n}(x - \bar{x})^2 = 172\cdot875$. Thus the variance of *b* is $1\cdot8053/172\cdot875 = 0\cdot010\,44$, and its square root is $0\cdot102$. The d.f. for $\hat{\sigma}^2$ are 6; the values of $t_{(6)}$ at 5% and 1% are respectively 2·447 and 3·707. Therefore 95% limits for *b* are $3\cdot94 \pm 2\cdot447 \times 0\cdot102$, i.e. $3\cdot94 \pm 0\cdot25$; so with 95% confidence we may say that the true value of the slope of the regression line relating *y* to *x* is not less than 3·69 nor greater than 4·19. To obtain 99% limits, 2·447 is replaced by 3·707 in the above calculation. Confidence limits to calculated values $y_{i.\text{Pred}}$ are not so easy to give, and a more advanced text should be consulted if these are needed.

14.4 Non-linear components of regression

The F-test above is really testing whether the regression-mean-square is considerably larger than the deviations-mean-square. We use the deviations-means-square as the estimate of σ^2, but strictly speaking it is not the best estimate that could be found. If it is possible to read duplicate values of *y* at each value of *x*, we can obtain an estimate of $\hat{\sigma}^2$ that genuinely measures only the random variation in *y*. In practice, this is not usually easy, or even possible, to arrange, and then we must rely on the deviations from linearity for a measure of σ^2. However, this ignores the possibility of there being a slightly systematic trend in the deviations, so small that the regression-mean-square is still much greater than the actual size of the deviations, but none the less apparent from a graph like Fig. 14.2. Here a straight line is, statistically, a very good fit; but the deviations from it are *systematic* and not random. No statistical test can make up for the failure to examine the data carefully before fitting a line. Computer graphics make this easy to do. The best we can do in a numerical analysis is to test whether a line *fits* well, not whether it provides a reasonable hypothesis. The reason we take as Null Hypothesis for a significance test the value $b = 0$ is that if there is no relation at all between *y* and *x*, then *y* is simply varying randomly about its mean $\bar{y}$, and is neither increasing (*b* positive) nor decreasing (*b* negative) with changes in *x*.

When it is clear that linear regression is not a satisfactory explanation of the relation

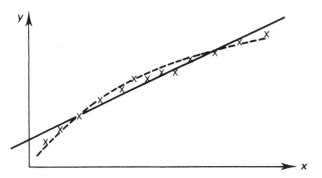

Fig. 14.2 Slightly curvilinear regression relation between y and x.

between x and y (either from inspection of the graph or because $F_{(1,\ n-2)}$ is not significant), another useful approach for the biologist is to look for a *logarithmic* relation. If a curve like Fig. 14.3(a) appears, then $\log y = a + bx$ may fit, or if the curve is like Fig. 14.3(b) then it is worth examining $y = a + b \log x$. Thus if the original data seem to follow either of these two curves, a straight line may be obtained by plotting (a) y on a logarithmic scale and x on an ordinary one, or (b) y on an ordinary scale and x on a logarithmic one. To fit a line, the calculations proceed as described above except, of course, that in case (a) each value of y_i is replaced by its logarithm while x_i remains in its original units; or in case (b), y_i is unchanged and x_i is replaced by $\log x_i$. There is nothing unnatural in using logarithmic units when necessary: indeed, Fig. 14.3(a) is often relevant where y is the size of a cell or culture and x is the amount of an added growth-stimulant substance present in the medium in which the cell or culture is growing. A similar growth curve also occurs, with x in this case measuring time, if the *rate* of growth of a cell or organism at any instant of time is proportional to its actual size at that instant. These curves are usually called *exponential*; the relation $\log_e y = a + bx$ is mathematically equivalent to $y = ce^{kx}$, in which c, k are two other constants replacing a, b and e^x is the exponential function that was employed in Chapters 6 and 7. (See Causton[6] for further explanation.) An example of Fig. 14.3(b) arises in insecticide and fungicide work, wherein the logarithm of the concentration, x, of applied insecticide or fungicide is often found to be linearly related to the numbers of insects or spores killed, y.

It is sometimes advocated that when $y = a + bx$ clearly does not fit, and the relation is obviously a curved one, a polynomial in x should be developed,

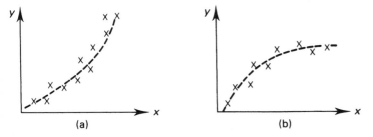

Fig. 14.3 (a) Relation between x, y which leads to the regression line $\log y = a + bx$; (b) relation leading to $y = a + b \log x$.

$y = a + bx + cx^2 + dx^3 + \cdots$, higher powers of x continuing to be added until a sufficiently close fit is found. Although the arithmetical difficulty of fitting has been removed by the availability of statistical computing packages, it is usually an insuperable problem to explain the result at all sensibly; and where regression is being used to predict the behaviour of y in relation to x, on the basis of a number of observations, n, which may be quite small, this is very dangerous. It is even more dangerous to predict what might happen outside the range of the observations actually taken: for suppose that in Fig. 14.3(a) we had lacked the two points on the graph corresponding to the two lowest x-values. Without them, no strong curvilinear tendency would be apparent, and so, no doubt, a straight line would be fitted. This would cross the x-axis to the right of 0, and we might therefore waste much time searching for a reason why y should be zero while x is positive; but in fact this value of x would lie outside the range of x-values available when the line was being determined.

14.5 Which regression?

Sometimes it is not clear at once which of x, y is dependent on the other: a graph may look more like Fig. 13.1(a) or (d) than Fig. 14.1. A classical example is to let one variate represent the height of a father, and the other variate the height of his son. Can any meaning be attached to 'the regression of sons' heights on fathers' heights' or 'the regression of fathers' heights on sons' heights', and if so which is the more fundamental? Obviously sons' heights depend on fathers', and not vice versa, so sons' heights would be taken as y; but x cannot now be thought of as *fixed* in the same sense as in Example 14.1.1. So long as x, the fathers' heights, can be measured with errors which are negligible in relation to the heights being considered, this will not matter, and $y = a + bx$ can be fitted as above. But could there be any purpose in trying to fit $x = \alpha + \beta y$? As a method of predicting x_i, a father's height (presumed unknown), from y_i, the height of his son (presumed known), it might be useful. If, on these grounds, the calculation is carried out, the estimates $\hat{\alpha}$ and $\hat{\beta}$ will be those which make deviations from the line in the *horizontal* direction as small as possible, and the line $x = \hat{\alpha} + \hat{\beta} y$ will thus be different, perhaps considerably different, from the line $y = \hat{a} + \hat{b} x$ because $\hat{\beta}$ will have $\sum_{i=1}^{n} (y_i - \bar{y})^2$ in its denominator instead of the $\sum_{i=1}^{n} (x_i - \bar{x})^2$ appearing in $\hat{b}$. This emphasises the importance of getting the logic of dependence and independence right before making calculations. If there seems to be no convincing reason for taking either variate as the independent one, it is perhaps unwise to try fitting so specific a relation as linear regression, but to rely on a correlation coefficient as a measure of relationship only, without introducing the concept of dependence at all.

14.6 Regression in MINITAB

There is a command REGRESS which carries out all the operations described in this chapter. The two variables x and y have to be named, e.g. 'conc' and 'optical' in Example 14.1.1. The x-variable (conc) is called the *predictor*, and the basic instruction is

REGRESS 'OPTICAL' on 1 predictor, 'CONC'.

For simple linear regression we have to say *1 predictor* because this is a special case of the general problem of fitting $y = a + b_1 x_1 + b_2 x_2 + \cdots$ where there may be several xs which together help to predict the value of y.

Output includes the equation, the standard deviation of $\hat{a}$ and $\hat{b}$, a *t*-ratio for each of $\hat{a}$ and $\hat{b}$ which allows us to test the Null Hypotheses '$a=0$' and '$b=0$', an estimate of σ^2 and the Analysis of Variance. There is also R^2, the square of the correlation coefficient between x and y, which may be used to show how well a straight line fits the observed data. It is dangerous to rely too much on R^2 however, because it cannot distinguish a random scatter of points about the line, such as that shown in Fig. 14.1, from a more systematic pattern of deviations like Fig. 14.2. There is no substitute for drawing a diagram to obtain proper interpretation of calculated results.

Additional information can be found using the PREDICT subcommand, which gives the *y*-value on the fitted line for any specified x, together with confidence intervals both for the mean of y and for individual values of y to be expected at that value of x.

14.7 Exercises on Chapter 14

14.7.1   Six fertiliser treatments were applied to plots of sugar beet, and the crop yield was recorded for each. The treatments differed only in the amount of fertiliser applied, not in its constitution. Calculate the regression line of y on x, test its goodness-of-fit, and

Treatment	(1)	(2)	(3)	(4)	(5)	(6)
Amount (cwt/acre) x	$\frac{1}{2}$	1	2	3	4	6
Yield (lb/plot) y	10	16	26	35	50	72

and find the predicted yields corresponding to each value of x used in the experiment and also to $x=5$ and $x=0$.

Comment on the latter.

Obtain also 95% confidence limits for the slope b of the line.

14.7.2   Data on the relationship between the chemical content of a particular constituent in a solution (g/l) and the crystallisation temperature (°K) were as follows:

Temperature	0·3	0·4	1·2	2·3	3·1	4·2	5·3
Content	3·2	2·4	4·3	5·4	6·6	7·8	8·8

Estimate the linear regression of content on temperature. Test whether a straight line is an adequate fit, and set 95% confidence limits to the true value of b, the regression coefficient.

14.7.3   The following data represent the size of an organism at equally spaced times 0 to 8. Plot these on a graph and consider how one might proceed to find a regression relation between size (y) and time (x).

x:	0	1	2	3	4	5	6	7	8
y:	0·75	1·20	1·75	2·50	3·45	4·70	6·20	8·25	11·5

Calculate the equation of a suitable linear regression function.

14.7.4 The weekly salaries of seven scientists of different ages in a department are as shown below. Calculate the linear regression equation of y on x. What would happen to the slope and intercept if (a) all the scientists had a £10-per-week increase, (b) they all had a 9% increase of salary?

Age, x years	20	22	28	33	36	45	55
Salary, £ y	69·8	78·5	108·8	134·0	144·0	182·5	198·6

15 The principles of experimental design: the completely randomised design

15.1 Designed experiments

In research, comparisons must often be made between several sets of data collected from basically similar populations, such as when groups of plants of the same type have been grown under conditions alike except that different fertilisers were used for each group. We saw in Chapter 8 that, on suitable assumptions, *two* such sets of data may be compared by a *t*-test; and this needs extension for more than two sets. There are also considerations (practical as well as theoretical) about how to collect the data used for making comparisons, and these two problems give rise to a science of *Experimental Design and Analysis*.

In a classical example, suppose that a field has been sown as uniformly as possible with a standard variety of wheat, and has been marked off into a number, N, of **plots** of constant size. These plots are the units of experimental material. Several different fertiliser treatments are being compared; the treatments are usually labelled by capital letters A, B, C, D, ... At the end of the season, the yield y of crop from each of the N plots is recorded. Consider what sources of variation there might be among these N different values of y; as we saw in Chapter 1, there will be natural experimental error variation from plot to plot even when no different treatments are applied. But to this *random* source of variation must be added a second *systematic* source, due to the different effects of the treatments: suppose, for example, that A is a better-balanced fertiliser than B, but not so good as C which supplies the nutrient elements to the plant in a more easily available form in the soil. In this case, all the yields on plots receiving treatment C would tend to be higher than those with A, which in turn would be higher than those with B; though within each treatment natural variation will also be apparent.

In this example, as in many others, the residual (experimental-error) distribution may be taken as normal (Chapter 7). Thus the yield distribution within any of the treatments has the same variance, σ^2, but the mean yields will be different; σ^2 is a characteristic of the type of plant which we are using but μ will vary depending on the fertiliser treatment. If μ_i is the mean yield per plot of the plants receiving treatment i, the distribution of plot yields for this treatment will be $N(\mu_i, \sigma^2)$. We must recognise, however, that μ_i is not a constant in any absolute sense, for when any of the treatments is repeated on another site or in a subsequent season, under soil and climatic conditions which are inevitably not quite the same, the mean yield under that treatment will alter—sometimes considerably. But what is true of treatment A, say, will equally be true of B and of all the rest: all will react to the environmental change to much the same extent, so that the *differences between* pairs of treatments will be much more nearly constant, and independent of the exact conditions of the experiment. The statistical analysis therefore concerns itself with comparing, rather than simply estimating, the mean yields under different treatments.

The methods about to be developed have applications to a vast number of situations in biology, industry, medicine and the physical sciences. In animal nutrition, various diets will be under test on animals bred under uniform and controlled conditions; in the textile industry, various different combinations of temperature and humidity during the weaving of a cloth will constitute the treatments being compared; in analytical chemistry, different operators doing the same analysis will form the treatments, which here measure personal differences—if these are small, the method, being reproducible, will be more useful than one which permits larger operator-to-operator variation; in metallurgy, specimen components made from batches of alloy produced by different suppliers will be compared for strength, with a view to seeing if all suppliers produce equally good material. *Treatments*, in the general sense, just imply a source of systematic variation.

Any experiment to compare several treatments must embody at least the first two *Principles of Experimental Design*.

1 Since there is variation from plot to plot, the results from several unit plots must be used in assessing the response to each treatment: this is *Replication* of treatments, and besides giving satisfactory estimates of μ_i it enables σ^2 to be estimated from the variation among plots treated alike.

2 The treatment means will be genuinely unbiased only if no conscious allocation of better or poorer plots to the various treatments is made; this demands *Randomisation*, i.e. random allocation of treatments so that each plot has the same chance of carrying any one of the treatments A, B, C, . . . which are under test. Another way of saying this, in the language of Chapter 4, is that the selection of plots to carry each treatment is a random sample from all those plots available.

15.2 The model for explaining the data

The simplest experimental design, which incorporates these two principles only, is the *Completely Randomised* layout. Suppose the treatments, called A, B, C, D, . . . , are t in number, and that each one is replicated r times, the total number of experimental plots therefore being $rt = N$. Unless some treatments are of considerably more interest than others, it seems intuitively sensible to replicate each the same number of times, and the analysis is simple when this is done. However, some situations require unequal replication and an example of this is given in Section 15.7. As a basis for the analysis of variance, set up the following model to explain how the yield (or whatever measurement has been taken per plot) arises:

$$y_{ij} = \mu + t_i + e_{ij} \tag{1}$$

In this, y_{ij} is the yield on the jth of the plots which carry treatment i (so that j takes values 1 to r), μ is the grand mean (average) yield over all the N plots, t_i is an *effect* due to treatment i (so that i takes values 1 to t), and e_{ij} is the natural residual variation or *experimental error* term on that particular plot. The effect t_i is a *deviation* from μ, positive for a better-than-average treatment, negative for a worse-than-average one; but since all these deviations are measured about the grand mean for the whole experiment they must themselves have mean 0, in other words their total $\sum_{i=1}^{t} t_i = 0$.

Whenever an Analysis of Variance is used, the error term e_{ij} must satisfy two conditions: (*i*) each e_{ij} is $N(0, \sigma^2)$, (*ii*) all e_{ij}s are mutually independent. Condition (*i*) implies that we must use material which, in the absence of different treatments, would give a normal distribution of yields, and also that the variance is the same for every treatment; this can reasonably be assumed unless some of the treatments have severe or unusual effects.

Condition (*ii*) is satisfied if the layout of treatments on plots really is random: if not, the yields and experimental errors on adjacent plots are likely to be correlated to some extent. Furthermore, the so-called *additivity* condition must hold: the mathematical model above must really consist of parts added together, not multiplied or combined in any other way. In Chapter 18 we consider how to proceed if some of these conditions do not hold.

15.3 The analysis of variance

This technique has two purposes: to find a valid estimate of σ^2 and to compare the mean yields under treatments A, B, C, D, ... A Total Sum of Squares S is defined, in the same way as in Chapter 14, to be the sum of all the squared deviations of individual plot yields from their overall mean, so that

$$S = \sum_{i=1}^{t} \sum_{j=1}^{r} (y_{ij} - \bar{y})^2.$$

The symbol $\sum_{i=1}^{t} \sum_{j=1}^{r}$ implies that we add up all the squares obtained by letting both suffices i and j run through all their possible values, from 1 to t and 1 to r respectively; one $\sum$ is written for each suffix. In practical terms this means that we take the sum of squares over all the N plots in the experiment. This total sum of squares is now split into two parts, one corresponding to each of the terms in Model (1) of Section 15.2, namely treatments and residual; note that μ has already been accounted for by measuring S about the grand mean $\bar{y}$.

As usual, let $G =$ grand total of all experimental observations, so that in this case

$$G = \sum_{i=1}^{t} \sum_{j=1}^{r} y_{ij},$$

that is the sum of yields over all the $N\ (=rt)$ plots. The grand mean yield in the whole experiment is $\bar{y} = G/N$ (or G/rt). If we use the symbol T_i to stand for the total yield of all the r plots which carried the treatment i, then

$$T_i = \sum_{j=1}^{r} y_{ij}$$

and the mean yield of all plots receiving treatment i is T_i/r. This is what we referred to as μ_i on page 139, but in Model (1) of Section 15.2 we found it more useful to call it $\mu + t_i$. A *Sum of Squares for Treatments* is defined as

$$S_T = \sum_{i=1}^{t} \sum_{j=1}^{r} \left(\frac{T_i}{r} - \bar{y} \right)^2$$

using the deviations of the treatment means from the grand mean. When all treatment means are of similar size, and so all near in value to $\bar{y}$, then S_T will be small, but if S_T is large this implies that some means will be much larger than others and the whole set will need further examination. The *Residual Sum of Squares*, S_E is the difference between S and S_T: $S_E = S - S_T$; we shall always compute S_E by taking this difference (or an equivalent one in other designs, as in Chapter 16) and never by developing a special formula for it. However, the formulae for S, S_T are easy to express in terms of the y_{ij} and T_i,

$$S = \sum_{i=1}^{t} \sum_{j=1}^{r} (y_{ij})^2 - \frac{G^2}{N}$$

$$S_T = \frac{1}{r} \sum_{i=1}^{t} (T_i)^2 - \frac{G^2}{N}.$$

In the problems we usually meet, where N is not very large, these formulae are simple and accurate to use; but they should not be used when programming for electronic computers.

15.3.1 Example
Three fertiliser treatments, A, B, C, each applied to seven plots of strawberry plants resulted in the following weights of crop (lb/plot):

A:	24,	18,	18,	29,	22,	17,	15;	total 143
B:	46,	39,	37,	50,	44,	45,	30;	total 291
C:	32,	30,	26,	41,	36,	28,	27;	total 220

The treatment totals ($T_A = 143$, $T_B = 291$, $T_C = 220$) add to give the grand total of all crop weights in the experiment, namely $G = 654$. There were three treatments each replicated seven times, i.e. $t = 3$ and $r = 7$; so $N = rt = 21$. Thus

$$S = (24^2 + 18^2 + 18^2 + 29^2 + 22^2 + 17^2 + 15^2 + 46^2 + 39^2 + 37^2 + 50^2$$
$$+ 44^2 + 45^2 + 30^2 + 32^2 + 30^2 + 26^2 + 41^2 + 36^2 + 28^2 + 27^2)$$
$$- \frac{654^2}{21}$$
$$= 22\,520 - \frac{427\,716}{21} = 22\,520 - 20\,367 \cdot 43 = 2152 \cdot 57$$

Also

$$S_T = \frac{1}{7}(143^2 + 291^2 + 220^2) - \frac{(654)^2}{21} = \frac{153\ 530}{7} - 20\ 367 \cdot 43$$

$$= 21\ 932 \cdot 86 - 20\ 367 \cdot 43 = 1565 \cdot 43$$

Hence $S_E = S - S_T = 587 \cdot 14$. Table 15.1 sets out the Analysis of Variance in symbolic form, and Table 15.2 the calculations for Example 15.3.1. Asterisks denote the level of significance, as on p. 133.

Table 15.1 Analysis of Variance for a completely randomised design

Source of variation	D.F.	Sum of squares	Mean square	
Treatments	$t-1$	S_T	$S_T/(t-1)$	$\dfrac{\text{Trt.M.S.}}{\text{Residual M.S.}} = F_{[(t-1),\ t(r-1)]}$
Residual	$t(r-1)$	S_E	$S_E/[t(r-1)]$	
Total	$rt-1$	S		

Table 15.2 Analysis of Variance for the data of Example 15.3.1

Source of variation	D.F.	Sum of squares	Mean square	F-ratio
Treatments	2	1565·43	782·72	$F_{(2,\ 18)} = 24 \cdot 00$***
Residual	18	587·14	32·62	
Total	20	2152·57		

The value of σ^2 is estimated to be 32·62. The Null Hypothesis, of no effect of treatments, is rejected at the 0·1% significance level.

Before we can proceed further with the Analysis of Variance, we must consider exactly what hypotheses we are testing statistically. Model (1) has already been set up (page 140); after having studied Chapter 8 readers will realise that this Model is in a form suitable to serve as an Alternative Hypothesis, which we shall accept only when we can show that some simpler Null Hypothesis is unreasonable. There is only one simpler idea that we can set up that is relevant to the present analysis, namely that the t_i can be omitted. So the Null Hypothesis will be

$$y_{ij} = \mu + e_{ij} \qquad (2)$$

This contains the random term only, and states exactly what we have assumed to be true if no effectively different treatments had been applied, namely that every y_{ij} would be $N(\mu, \sigma^2)$. We now consider what can be said about S_T and S_E on this Null Hypothesis: they will both be sums of squares of normal deviates, so having χ^2 distributions (Chapter 11), with degrees of freedom $(t-1)$ and $t(r-1)$ respectively (the latter being equal to $(N-t)$, i.e. $rt-t$).

There is a simple working rule for obtaining degrees of freedom: there are t treatments, and so it is possible to compare 1 with 2 (by calculating $t_2 - t_1$), 1 with 3, 1 with 4 and so on up to 1 with t, all of these comparisons, $(t-1)$ in number, being independent of one another. But when we try to compare 2 with 3, this gives no fresh or independent information since (t_3-t_2) is just $(t_3-t_1)-(t_2-t_1)$ and these two comparisons have already been made; this is equally true for any other comparison attempted after the first $(t-1)$. So the degrees of freedom, the number of independent comparisons possible among the treatments, equal $(t-1)$, *one less than the actual number of treatments.* Similarly, among all the rt observations y_{ij}, $(rt-1)$ independent comparisons can be made, so that S has d.f. $(rt-1)$. As S_E is the difference between S and S_T so its d.f. are (d.f. of S) *minus* (d.f. of S_T); residual degrees of freedom in more complicated types of experimental design are always calculated by taking away from the d.f. of S all the d.f. of the other sums of squares contained in S (see Chapter 16).

As we saw in Chapter 14, the *mean squares* for each term in the Analysis of Variance are useful: these are the sums of squares divided by their corresponding degrees of freedom. Since S_T is $\chi^2_{(t-1)}$, on the Null Hypothesis, $S_T/(t-1)$ is an estimate of σ^2; similarly $S_E/[t(r-1)]$ also estimates σ^2. Therefore, on the Null Hypothesis the ratio *treatments-mean-square/residual-mean-square* is distributed as $F_{[(t-1),\ t(r-1)]}$ since it is the ratio of two expressions each of which estimates σ^2. Hence (Chapter 11) the numerical value of this ratio should not differ significantly from 1, and if it does so we shall reject the Null Hypothesis upon which this theory has been built. That is to say, the simple model $y_{ij}=\mu+e_{ij}$ will not be satisfactory. Having rejected the Null Hypothesis, we accept the Alternative Hypothesis, Model (1). This process is automatic, for we never actually test anything about the Alternative. But, in fact, it is a reasonable one for situations where the F-ratio *does* exceed 1, because we can show that on the Alternative Hypothesis the treatments-mean-square estimates $(\sigma^2+r\sum_{i=1}^{t}t_i^2/(t-1))$, and this must exceed σ^2 since the t_i are not all zero (as they are on the Null Hypothesis). The residual mean square still estimates σ^2; it therefore gives us the necessary estimate of σ^2 whichever hypothesis we decide is appropriate.

At the outset, we aimed to estimate σ^2, which we have now done, and also to compare the various treatments. So far, we have seen only whether the treatments can be considered a homogeneous set or not: if not (i.e. when F is significant), we want to look at the so-called *treatment means*—the mean yields under the various treatments—and especially, as we saw earlier, at their differences. For *any pair* of means, a test of the significance of their difference is provided by the t-test (Test 8.7.3), since each mean has been formed from r observations from a normal distribution whose variance is estimated by $\hat{\sigma}^2$. So for A and B

$$\frac{T_A/r - T_B/r}{\sqrt{2\hat{\sigma}^2/r}}$$

is distributed as $t_{(f)}$, where f denotes the residual d.f.; f is equal to $t(r-1)$ in the completely randomised design. Writing $t_{(0.05,\ f)}$ to stand for the value shown in the t-table (Table I) at the 5% significance level and with f degrees of freedom, we see at once that if

$$\frac{T_A/r - T_B/r}{\sqrt{2\hat{\sigma}^2/r}}$$

exceeds $t_{(0.05,\ f)}$, the two means differ significantly at 5%. This is exactly the same as to say that there is significance when $T_A/r - T_B/r$ exceeds $(t_{(0.05, f)} \times \sqrt{2\hat{\sigma}^2/r})$, the latter expression being called the **significant difference** (sometimes *least significant difference*) between the two means. When all the treatments have, in the simplest case, been replicated equally (r times) the same calculation is required for testing significance between *any* two means; therefore the final step in an analysis wherein F has proved significant is to work out a significant difference (usually at each of the significance levels 5%, 1% and 0·1%) and use it to compare the treatment means. Any pair of means whose difference is greater than the significant difference may be declared significantly different.

15.3.1 Example (concluded)

The treatment means are, for A, $143/7 = 20·43$, for B, $291/7 = 41·57$, for C, $220/7 = 31·43$. The residual-mean-square estimate of σ^2 is $\hat{\sigma}^2 = 32·62$. The residual d.f, $f = 18$, and $t_{(0.05, 18)} = 2·101$. The expression $\sqrt{2\hat{\sigma}^2/r}$ is $\sqrt{2 \times 32·62/7} = \sqrt{9·32} = 3·05$, giving a significant difference of $2·101 \times 3·05 = 6·41$ at the 5% level. Similarly $t_{(0.01, 18)} = 2·878$, leading to a significant difference of 8·78, and $t_{(0.001, 18)} = 3·922$, leading to a significant difference of 11·96. Now the means of A, C differ by 11·00, which is greater than 8·78 but not so great as 11·96; so A, C differ significantly at the 1% level but not at 0·1%. Likewise, A, B differ at 0·1%, and B, C at the 1% level.

15.4 Contrasts

The reader may have spotted an anomaly in the conclusion of Example 15.3.1, since there are only 2 d.f. for treatments and yet three comparisons have been made. There is, in fact, a greater fundamental difficulty even than this, for if we do make comparisons between all possible pairs of means we violate the level of significance (5% etc.) that we claim to be using, in the sense that our results are significant not at that level but at some much less worthwhile one. The actual size of level depends on how many treatments are included in the experiment and which pair we test. The unthinking practice of testing the largest treatment against the smallest can give some wildly wrong answers and, indeed, if the experiment contains enough treatments—even though these may form a homogeneous set—a *t*-test between the largest and smallest is highly likely to show a significant difference. So as to reduce the risk of making wrong statements after carrying out the analysis, various safeguards have been suggested.

Firstly, note that in Example 15.3.1 an F-test was carried out first, and only when F proved significant did we proceed further, to *t*-tests or significant differences. Provided this order is always observed, serious errors are less likely. But, even then, we cannot properly compare every pair of means, and certainly it is unwise to make a larger number of statements about treatment differences than there are degrees of freedom for treatments (for not all of these statements can be independent of each other, to give fresh information). And such comparisons as are tested should have been decided on *before* the experimental results are seen rather than be suggested by them: in other words, we had some definite ideas that we wished to examine when embarking on the experiment—this is certainly not asking too much of a properly planned piece of research. Often there will be a relatively small number of comparisons or **contrasts** that are of use or interest, for example: (1) one treatment may be a *control*, either untreated or having a well-known, established treatment applied to it (such as a standard fertiliser) and the only interest is to compare the other treatments with this one; (2) the treatments can be split up into groups, such as when a plant-breeder puts a large number of new crosses, raised at the same time, into the same experiment—the treatments here are the different crosses—but really wants to compare only crosses which have a parent in common; (3) the treatments form a *factorial* set, considered in Chapter 17. When only a small number of comparisons have been nominated for testing, it is correct to omit F-tests and proceed to *t*-tests at once, using the estimate of $\hat{\sigma}^2$ in the Analysis of Variance. In this situation, the sole reason for putting a mixed-looking set of treatments into the same experiment may be that it is convenient to have only one piece of work in progress on the same type of plant at the same time, and to obtain the estimate $\hat{\sigma}^2$ from as many plots as possible. If some more specific hypothesis about the relation of y to the treatments is possible, such as a regression (Chapter 14) or other form of response curve (Chapter 17, pp. 169–70), this will of course be tested in preference to making large numbers of individual treatment comparisons.

Contrasts is the name usually given to comparisons that may involve more than two means (see Example 15.7.1), and when an experiment has been properly planned to answer certain questions, some of which may involve groups of means, the overall F-test is irrelevant; it is these contrasts which have to be studied. A modern text on Experimental Design (e.g. Clarke and Kempson[7]) will explain this fully.

Secondly, significances or significant differences are sometimes deliberately not quoted in published results of experiments, preference being given to the *standard error of a treatment mean*, $\sqrt{\hat{\sigma}^2/r}$, or to the *standard error of a difference between two means*, $\sqrt{2\hat{\sigma}^2/r}$. Of course, in this as in all other methods, an Analysis of Variance table is prepared, because it is the best way of obtaining the necessary estimate of σ^2. But since comparisons will inevitably be made by people wishing to apply the results of these experiments, significant differences will, in practice, be calculated in any case, by multiplying the first of these standard errors by $(t_{(0.05, f)} \sqrt{2})$ or the second by $t_{(0.05, f)}$. However the results are presented, the temptation to extract too many comparisons still has to be avoided. And of course the remarks of Chapter 8 about the significance level used (5%, 1%, etc.) apply equally here.

Thirdly, there may be occasions when it really is useful to compare the whole set of treatments. Usually this will only be the right approach in the early stages of a programme of work, with a view to choosing what treatments should be included in follow-up studies—where some treatments will eventually be discarded. Statistical evidence of which seem the best and worst treatments can be a useful addition to other biological considerations in deciding what should be included in the next stage of the programme: in the plant-breeding example mentioned earlier, those high-yielding crosses which have good growth habit and are disease resistant will be examined further. The plant breeder will not look *only* at the statistics of yield, of course; he will use all his biological knowledge, and observe as well as measure properties that seem to him desirable. Various statistical tests have been proposed for carrying out indiscriminate comparisons of all treatments; *t*-tests should not be used for the reasons given above. One of the more commonly used is Duncan's test (called a 'multiple-range' test). First σ^2 is estimated in the usual way, then a list of the means is written down in increasing order of size, and the final aim is to form groups of means that do not differ from one another significantly. To do this, special tables are needed; these, and a detailed description of the method, may be found in Duncan's original paper[8] or in the book by Edwards.[9] The practice, sometimes adopted in journals, of using Duncan's test instead of *t*-tests when reporting the final results of a programme of planned experiments, is not correct, especially when the treatments form a carefully structured set such as a factorial (Chapter 17). Duncan's test, and others like it, apply at the stage in the programme where possible treatments are being explored and 'screened', not at the time when specific hypotheses have been set up for testing.

15.5 Layout of the experiment

To satisfy the principle of randomisation, we number the available field plots 1 to 21 and arrange that a random selection of seven of the numbers between 1 and 21 carries A, a further random seven, B, and the rest, C. In order to use a run of random digits from a table (such as Table VI), mark these off in pairs, for example 07, 43, 55, 27, 18, 34, 94, 56, ... Now 07 corresponds to plot number 7, so we write this down first in our list; then for numbers greater than 21, let 22–42 correspond respectively to 1–21, likewise 43–63 and 64–84 also correspond to 1–21, while 85 upwards (and 00) are ignored. Thus 43 corresponds to 1; 55 to 13; 27 to 6, 18 is itself; 34 corresponds to 13, but this has already appeared once; 94 is ignored; 56 corresponds to 14, and so on. Neglecting all repeats, we arrive by this process at a list of numbers, between 1 and 21, in random order, say: 7, 1, 13, 6, 18, 14, 2; 19, 3, 11, 21, 16, 17, 5; 8, 20, ... This may be called a *random permutation* of the numbers 1–21, and such permutations are printed by Fisher and Yates[2] and Cochran and Cox[10], though not for numbers above 20. The first seven plots in the list are allocated to A, the next seven to B, and the rest to C, as in Fig. 15.1.

1	2	3	4	5	6	7
A	A	B	C	B	A	A
8	**9**	**10**	**11**	**12**	**13**	**14**
C	C	C	B	C	A	A
15	**16**	**17**	**18**	**19**	**20**	**21**
C	B	B	A	B	C	B

Fig. 15.1 Completely randomised layout for seven replicates of three treatments.

This layout may look rather systematic, to the extent that the treatments are somewhat bunched, especially A and C: had we been producing a layout 'haphazardly, out of the head', we would hardly have accepted this one. But it is a perfectly valid one, satisfying both principles so far laid down. If there are valid reasons for suspecting systematic soil variation, so that we prefer a more even scatter of treatments through the field, the third Principle, that of *Blocking*, is used; wherever possible the layout will also satisfy the fourth Principle, *Orthogonality*. These two further principles are set out in the next chapter.

15.6 Assumptions in the Analysis of Variance

There are three basic assumptions which must hold, at least to a very good degree of approximation, if the Analysis of Variance technique is to be valid. It is used so often that perhaps these are taken too much for granted.

First (see page 140), the Model set up to explain the observations—Model (1) and others like it in later chapters—must be *additive*. Second, the measurement which we analyse must be *normally* distributed, or very nearly so; a skew distribution for y would make the results very unreliable. Third, σ^2 must be *constant* for all plots, that is to say, also, for all treatments. It is not easy to test for the first two of these conditions, though in Chapter 18 we consider practical ways of analysing non-normal data. We also consider non-constant variance in Chapter 18, but for this there *is* a test available.

Bartlett's test for homogeneity of variance takes k sets of data, from each of which an estimate of variance has been calculated; the ith estimate is s_i^2, on f_i degrees of freedom. First compute an 'average' variance

$$s^{*2} = \sum_{i=1}^{k} f_i s_i^2 \bigg/ \sum_{i=1}^{k} f_i.$$

Then obtain

$$M = \left(\sum_{i=1}^{k} f_i \right) \log_e s^{*2} - \sum_{i=1}^{k} (f_i \log_e s_i^2)$$

and

$$C = 1 + \frac{1}{3(k-1)} \left(\sum_{i=1}^{k} \left(\frac{1}{f_i} \right) - 1 \bigg/ \sum_{i=1}^{k} f_i \right).$$

Finally M/C is tested as $\chi^2_{(k-1)}$.

15.6.1 Example
In four sets of data, estimated variances are 14·5 (on 8 d.f.), 35·4 (6 d.f.), 67·2 (23 d.f.) and 23·3 (12 d.f.). Do these differ significantly?

$s^{*2} = ((8 \times 14\cdot5) + (6 \times 35\cdot4) + (23 \times 67\cdot2) + (12 \times 23\cdot3))/(8 + 6 + 23 + 12)$

$\quad = 2153\cdot6/49 = 43\cdot9510$

$M = (49 \times 3\cdot7831) - \{(8 \times 2\cdot6741) + (6 \times 3\cdot5667) + (23 \times 4\cdot2077) + (12 \times 3\cdot1485)\}$

$\quad = 8\cdot0193.$

$C = 1 + \frac{1}{9}(\frac{1}{8} + \frac{1}{6} + \frac{1}{23} + \frac{1}{12} - \frac{1}{49}) = 1\cdot0442$, so $M/C = 7\cdot68$ which is not quite significant when tested as $\chi^2_{(3)}$, so that we cannot reject the Null Hypothesis that all variances are the same. Unfortunately, variances do have to be very different before the test shows significance, and non-normality of data has a serious effect on the test. Exercises 18.4 and 18.5 use this method. (Note that if logs to base 10, rather than base e, are used, the expression given in the formula for M must be multiplied by 2·3026.)

15.7 Unequal replication of treatments

15.7.1 Example
There is no need for every treatment to have the same replication. Sometimes not all the unit plots survive, or sometimes it is important to estimate the mean of one

treatment more precisely than another. In a completely randomised layout the analysis is no more difficult with unequal replication than with equal replication. Unfortunately we shall find this is no longer true for more complicated designs! The only difference in a completely randomised design lies in the way we compute the sum of squares for Treatments. When Treatment i is replicated r_i times, giving a total T_i, we require T_i^2/r_i. The grand total G is as usual $\sum T_i$ and the total number of units N is $\sum r_i$. The treatment sum of squares is now computed as $\sum T_i^2/r_i - G^2/N$ and the total sum of squares in the same way as usual. The residual sum of squares is the difference between these two. The rules for degrees of freedom also follow in the normal way: total will be $(N-1)$ because there are altogether N observations, and treatments will have one less than the number of treatments present in the experiment. Residual d.f. will be the difference between these two.

Data on the heights (m) grown by the three species of tree in a forestry plantation were:

Species	Code	r_i		Total T_i
Pinus caribea	C	9	4·2, 4·3, 3·5, 3·9, 5·0, 4·8, 4·6, 4·5, 4·0	38·80
Pinus kesiya	K	12	3·95, 3.85, 4·25, 4·7. 4·15, 3·3, 3·65, 3·7, 3·95, 4·0, 3·7, 4·3	47·50
Eucalyptus deglupta	E	8	7·95, 8·1, 8·3, 6·6, 7·5, 7·7, 7·25, 8·0	61·40
Total		$N = 29$		$G = 147·70$

The sum of squares for treatments is $826·5369 - 752·2514 = 74·2855$.
The sum of squares of all the observations is $831·8900$, and therefore the total sum of squares is $831·8900 - 752·2514 = 79·6386$.

Analysis of Variance

	D.F.	S.S.	M.S.	
Species	2	74·2855	37·1428	$F_{(2, 26)} = 180·39$***
Residual	26	5·3531	$0·2059 = s^2$	
Total	28	79·6386		

The F-value is very large indeed, and if it is of any interest to test the Null Hypothesis of no difference in growth between the species then clearly this N.H. is rejected. Comparisons among means will be made according to what it makes biological sense to report on.

The two *Pinus* species may be compared: the means are $C = \frac{38·8}{9} = 4·311$, $K = \frac{47·5}{12} = 3·958$. The variance of the difference is

$$\frac{s^2}{9} + \frac{s^2}{12} = 0·2059 \left(\frac{1}{9} + \frac{1}{12}\right) = 0·040\,04.$$

For a test of the N.H. 'mean of C = mean of K', $(\bar{y}_c - \bar{y}_k)/\sqrt{0·04\,004}$ follows the $t_{(26)}$ distribution (degrees of freedom are always those of s^2). $\frac{(4·311-3·958)}{\sqrt{0·04\,004}} = \frac{0·353}{0·200} = 1·77$, which is well within the central 95% of $t_{(26)}$ and so we do not reject this N.H.

An alternative procedure is to find a 95% confidence interval for $(\mu_C - \mu_K)$. The critical 5% value of $t_{(26)}$ is 2·056, and so the rule 'observed difference plus or minus t

times its standard error' gives $0.353 \pm 2.056 \times 0.200$, or 0.353 ± 0.411, which is (-0.058 to $+0.764$).

Another comparison or **contrast** we might have proposed (*before* looking at the data) would be E against all of the *Pinus*. The mean of E is $\frac{61.4}{8} = 7.675$, and of C and K together is $\frac{(38.8+47.5)}{(9+12)} = \frac{86.3}{21} = 4.110$; and the variance of the difference E *minus* (C and K) is $s^2 \left(\frac{1}{8} + \frac{1}{21}\right) = 0.0355$, the standard error being 0.189. A significance test gives $t_{(26)} = \frac{(7.675-4.110)}{0.189} = 18.86$ which is so large that a N.H. 'mean of Eucalyptus = mean of Pinus' would be totally untenable.

15.8 MINITAB One-way Analysis of Variance

A useful first step in comparing several sets of data is to show the dot-plots, one for each treatment, on the same scales beneath one another. This helps to check the basic assumptions that the data within each set are (approximately) normally distributed and have similar variances, so that comparisons based on a pooled variance are valid.

AOVONEWAY (so called because there is only one source of systematic variation, namely treatments) provides an analysis in the same form as Table 15.1. Treatments are called *Factors* and Residual is called *Error*. A list of the treatments (labelled by the columns in which their data are stored), and the number of observations for each, is given, with the mean for each treatment and the pooled variance. Each treatment variance is also given as a further check that they are not very different and can reasonably be pooled. A 95% confidence interval for each mean is shown, but is not of great use in studying contrasts. This program works for unequal, as well as equal, replication of treatments.

An extension called ONEWAY also allows *residuals* (the estimates of e_{ij} on each plot) to be output; sometimes a large residual, or several large residuals, will repay study by indicating some biological problem with that unit (plot, tree, animal etc.) or may suggest the need for further terms in the statistical model (see Clarke and Kempson[7]).

15.9 Exercises on Chapter 15

15.9.1 A sample of plant material is thoroughly mixed, and 15 aliquots taken from it for determination of potassium content. Three laboratory methods A, B, C are employed, A being the one generally used. Five aliquots are analysed by each method, giving the following results (μg/ml).

A:	1.83,	1.81,	1.84,	1.83,	1.79
B:	1.85,	1.82,	1.88,	1.86,	1.84
C:	1.80,	1.84,	1.80,	1.82,	1.79

Examine whether methods B and C give results comparable to those of method A.

15.9.2 Eight varieties, A–H, of blackcurrant cuttings are planted in square plots in a nursery, each plot containing the same number of cuttings. Four plots of each variety are planted, and the shoot length made in the first growing season is measured.
The plot totals (*m*) are:

A:	46,	29,	39,	35	E:	16,	37,	24,	30
B:	37,	31,	28,	44	F:	41,	28,	38,	29
C:	38,	50,	32,	36	G:	56,	48,	44,	44
D:	34,	19,	29,	41	H:	23,	31,	29,	37

B and C are standard varieties; assess the remaining six for vigour in comparison with B and C.

15.9.3 Strawberry plants were grown to provide fruit for comparing different preservation methods. Twice as many 'controls' C (a standard fertiliser) were grown as of any of the other four treatments (modified fertilisers) P, Q, R, S. Yields per plot are given to the nearest 10g unit.

C:	213, 274, 199, 313, 287, 205, 360, 252, 263, 184, 241, 226
P:	353, 271, 246, 318, 292, 387
Q:	408, 363, 359, 377, 284, 326
R:	225, 336, 291, 318, 357, 312
S:	462, 325, 388, 400, 379, 333

Examine the difference between C and each of P, Q, R, S. Also find a 95% confidence interval for the difference in mean yields of R and S.

15.9.4 Three culture media, A, B, C, are being used for growing colonies of fungal mycelium, and are being compared with a standard method, O, which is replicated 10 times while A, B, C are replicated only five times each. After 48 hours' incubation, the radii of the growing colonies are as follows (in units of the graduations in the microscope used to examine them):

A:	25,	23,	20,	24,	27					
B:	25,	28,	28,	24,	23					
C:	21,	18,	15,	19,	20					
O:	24,	21,	18,	20,	23,	21,	16,	19,	22,	17

Test whether any of A, B, C differ from the standard O.

16 Randomised blocks and the precision of results

16.1 Blocking

If it is known or suspected that there is another source of systematic variation, in addition to and independent of the treatments, an extra term is included in the Model, giving

$$y_{ij} = \mu + t_i + b_j + e_{ij} \tag{3}$$

In the example of a field experiment, this new source of variation will appear in the form of trends in fertility over the whole of the experimental area; if it is possible to find smaller areas each homogeneous within itself, these are used as *blocks*, and every block contains one complete set of the treatments, so that it must consist of t plots. Since there are to be r replicates of each treatment, there must be r blocks, and the suffix j in model (3) takes on a specific meaning: plot (ij) will now be that which carries treatment i in block j, and b_j is an *effect* due to blocks, having a similar character to t_i which measures treatments. The difference between blocks is made to account for as much as possible of the systematic soil variation, consistent with keeping block size at t plots. Figure 16.1 indicates how five treatments, each replicated four times, would be laid out in blocks to remove the effect of a fertility trend running from top to bottom of the diagram; following the usual convention we label the treatments by capital letters, and blocks in roman numerals.

With the trend shown, of soil conditions improving from the top to the bottom of Fig. 16.1, all plots in block I are naturally slightly inferior to those in block II, which in turn are inferior to those in block III, etc., so that in the absence of different treatments we would expect the yields in I to be systematically lower than those in II, and so on.

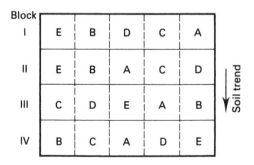

Fig. 16.1 Randomised block layout for four replicates of five treatments on land having a fertility trend in the direction shown.

Note that unless blocks run at *right angles* to the trend, all plots in each particular block are not equally affected by the trend, and b_j does not have the meaning intended. And finally, if one of the blocks is very different from the rest, its effect may not be independent of treatments: one very poor block of soil may induce a worse reaction to the less good fertilisers (in the language of Chapter 17, an *interaction* between blocks and treatments), and this invalidates Model (3).

Within each block, plots must be as closely comparable as possible (blocks are to be homogeneous within themselves) and so the blocks of Fig. 16.1 are subdivided in such a way as to make each plot equally affected by the trend. If the field is patchy, not showing a steady trend like this, then the blocks should be of such a shape as to remove as much patchiness as possible. The layout of treatments must be *random* within each block, a fresh randomisation being used for each block. For this purpose, tables of random permutations (referred to in Chapter 15) are often useful, but again it is possible to work directly from random numbers. We wish to order the digits 1, 2, 3, 4, 5 at random in four different ways, and then apply these to the treatments A, B, C, D, E in blocks I, II, III, IV. Waste of digits can be avoided by making 6 correspond to 1, 7 to 2, 8 to 3, 9 to 4, 0 to 5, so the run 07435527183494562314 becomes 52435522133444451 2314. The first four digits here are 5243, so giving 52431 as a random permutation of 1 to 5; then 552213 gives 52134; 344451 gives 34512; 2314 gives 23145. Translating to letters, we have 52431 giving EBDCA, and the others are EBACD, CDEAB and BCADE.

Blocks, like treatments, really mean only *a source of systematic variation*, and the name, though derived from the original application to field trials, may imply different batches of material in an industrial experiment, different litters of animals in a comparison of diets, or different machines or operators in a programme of analyses in a chemical laboratory. Even when there is no other reason for introducing blocks into a layout, they may still be put in to provide convenient units representing a day's work, or a morning's work or some other suitable amount of recording time: this helps if one person is not able to complete the job or if weather intervenes and makes the recording of a field experiment spread over a few sessions or a few days. In these cases, blocks may help to maintain homogeneity, and certainly will not reduce it.

16.2 Analysis

In the analysis, S and S_T are the same as for a completely randomised layout. We need to define $B_j=$ the total of all yields of plots in block j, so that the mean yield in block j is B_j/t (the block contains t plots). The total number of plots is $N=rt$. The blocks sum of squares is

$$S_B = \frac{1}{t}\sum_{j=1}^{r}(B_j)^2 - \frac{G^2}{N}, \text{ or } \frac{1}{N}\left(r\sum_{j=1}^{r}B_j^2 - G^2\right)$$

and will have $(r-1)$ degrees of freedom by the same argument as previously used. All the description in Chapter 15 of methods of testing for treatment differences can be carried through for block differences though, in the field, comparisons between block means are unlikely to be wanted: it will suffice to know that the blocks have done their job of removing some systematic variation. The residual sum of squares is $S_E = S - S_T - S_B$ and will have d.f. $(rt-1)-(t-1)-(r-1) = rt-r-t+1$, which simplifies to the product $(r-1)(t-1)$; as usual the residual mean square supplies the necessary estimate of σ^2. Table 16.1 gives the Analysis of Variance for a randomised block design.

Table 16.1   The Analysis of Variance for a randomised block design

Source of variation	D.F.	Sum of squares	Mean square	F-ratio
Blocks	$(r-1)$	S_B	$S_B/(r-1)$	Blocks M.S./Residual M.S. is $F_{[(r-1),(r-1)(t-1)]}$
Treatments	$(t-1)$	S_T	$S_T/(t-1)$	Trts. M.S./Residual M.S. is $F_{[(t-1),(r-1)(t-1)]}$
Residual	$(r-1)(t-1)$	$S_E = S - S_B - S_T$	$S_E/(r-1)(t-1)$	
Total	$(rt-1)$	S		

16.2.1 Example

Samples from five different suspensions of bacteria A, B, C, D, E, were examined under a microscope by four different observers, I, II, III, IV; the order in which each observer dealt with the samples was randomised to reduce errors due to fatigue, and the numbers of organisms recorded from the samples were as summarised in Table 16.2. (Note that if observer I worked in the order EBDCA, II worked EBACD, III worked CDEAB and IV worked BCADE, Fig. 16.1 would exactly represent the scheme followed—observers are *blocks*.)

Table 16.2   Data of Example 16.2.1

Observer number	Suspension					Observer's total
	A	B	C	D	E	
I	68	71	54	95	73	361
II	82	78	67	116	85	428
III	77	74	65	103	88	407
IV	59	70	54	90	76	349
Total	286	293	240	404	322	1545 = *Grand Total*

In this example, we may be interested in the actual comparison of observers, as well as in eliminating systematic differences due to them; and comparisons between certain of the suspensions will have been specified for testing. An Analysis of Variance is set up following the scheme of Table 16.1, and the details of this are given in Table 16.3.

$S_T = \frac{1}{4}(286^2 + 293^2 + 240^2 + 404^2 + 322^2) - \frac{1}{20}(1545^2)$, reducing to 3685·00;

$S_B = \frac{1}{5}(361^2 + 428^2 + 407^2 + 349^2) - \frac{1}{20}(1545^2)$, reducing to 839·75;

$S = (68^2 + 71^2 + \cdots + 90^2 + 76^2) - \frac{1}{20}(1545^2)$, the first term containing the squares of all the individual counts; S reduces to 4717·75.

Table 16.3   Analysis of data in Example 16.2.1

Source of variation	D.F.	S.S.	M.S.	F-tests
Blocks = Observers	3	839·75	279·917	$F_{(3, 12)} = 17·40***$
Treatments = Suspensions	4	3685·00	921·250	$F_{(4, 12)} = 57·28***$
Residual	12	193·00	16·083	
Total	19	4717·75		

Estimate of $\sigma^2 = 16·083$; the effects of Blocks and Treatments are both significant at the 0·1% level.

Since there are four observers, S_B has 3 d.f.; similarly S has 19 d.f. and S_T has 4 d.f. On subtracting, $S_E = 193·00$ and has 12 d.f.

Significant differences at the 5% level between means for observers (blocks) are $t_{(0·05, 12)}\sqrt{2\hat{\sigma}^2/5}$, because each observer-mean is based on five records and the estimate of σ^2 on 12 degrees of freedom; differences at the 1% and 0·1% levels may be calculated similarly. Means for suspensions are based on four records each, and so significant differences will be $t_{(0·05, 12)}\sqrt{2\hat{\sigma}^2/4}$ etc. Now $\sqrt{2 \times 16·083/5} = \sqrt{6·433} = 2·54$, and $\sqrt{2 \times 16·083/4} = \sqrt{8·042} = 2·84$, while the values of $t_{(12)}$ from tables are 2·179 (5%), 3·055 (1%) and 4·318 (0·1%). The complete list of significant differences for observers is thus $2·54 \times 2·179 = 5·53$ at 5%, $2·54 \times 3·055 = 7·76$ at 1% and $2·54 \times 4·318 = 10·97$ at 0·1%. Means for observers are: I, 72·2; II, 85·6; III, 81·4; IV, 69·8. Likewise, significant differences for suspensions are $2·84 \times 2·179 = 6·19$ at 5%, $2·84 \times 3·055 = 8·68$ at 1% and $2·84 \times 4·318 = 12·26$ at 0·1%. Means for suspensions are: A, 71·5; B, 73·3; C, 60·0; D, 101·0; E, 80·5.

After the discussion of Chapter 15, the reader will be rather wary of making all possible comparisons between the observer means or the suspension means. As regards suspensions, D is very much the greatest, significantly higher at 0·1% than any other. Also A and B are not very different from one another, while C appears to be lower than these two. If these results, or such parts of them as have been covered by the hypotheses to be tested, are sensible, they will be accepted with no further study; but if anomalies appear, such as if C had been expected to be very like D, the conduct of the experiment needs examination. The result of such examination is sometimes to discover sources of serious error, e.g. the suspensions not being properly shaken, and sometimes to reconsider the hypotheses. The comparison between observers in this example suggests that I and IV are different from II and III: explanations of these types of result would be sought in terms of skill, care or experience.

16.3 Orthogonality

This is not an essential principle of experimentation, but a very desirable one; the proper definition involves mathematical ideas outside our scope, but the implications of orthogonality are not hard to grasp. In randomised blocks, every block contains every treatment just once, so that if one block happens to be different from the others, every treatment is similarly affected. Had this particular block not contained every treatment, some adjustment would have had to be made to those treatments that were missing from it; it would not have been possible to consider treatment means by ignoring blocks—nor, indeed, block means by ignoring treatments, since the treatments absent from a particular block might, in their turn, have differed appreciably from those present in it. In an *orthogonal* design, each classification (blocks and treatments, also rows and columns in Latin Squares, below) can be examined independently of every other one.

16.4 Latin Squares

A randomised block removes one systematic source of variation additional to treatments; a Latin Square removes two such sources. In a field experiment, if two probable fertility trends can be thought of, in directions at right angles, both need to be made the basis of blocking. Thus when there is a slope in the land being used, and also a climatic trend (e.g. effects of wind, rain) at right angles to this, a randomised block cannot take out all the known variation. The Latin Square, previously studied by mathematicians purely for its pattern, has been applied to this situation. It is a layout in which every letter (A, B, C, ... , the set of treatments) occurs once in each row and once in each column, as illustrated in Fig. 16.2; this shows that the square must contain t rows and t columns, and hence t replicates of each treatment, and also $N = t^2$.

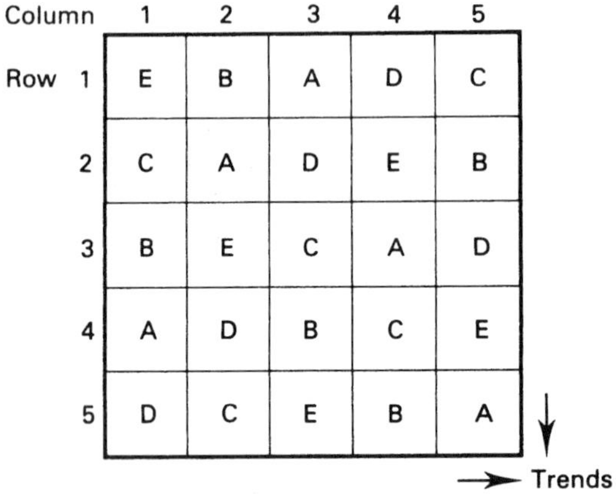

Column	1	2	3	4	5
Row 1	E	B	A	D	C
2	C	A	D	E	B
3	B	E	C	A	D
4	A	D	B	C	E
5	D	C	E	B	A

→ Trends

Fig. 16.2 Latin Square layout for five treatments.

Analysis proceeds along the same lines as for randomised blocks, but instead of the one sum of squares for blocks, systematic variation is now taken out by two sums of squares which are always called Rows (S_R) and Columns (S_C). The mathematical Model will now read

$$y_{ijk} = \mu + t_i + r_j + c_k + e_{ijk} \tag{4}$$

in which t_i refers to treatments, as usual, and the two terms r_j, c_k to rows and columns respectively in the layout; e_{ijk} is a residual error term with all the usual properties. The use of three suffices i, j, k, is sometimes a source of confusion, especially as each may take the values 1 to t; but a glance at Fig. 16.2 will make it clear that when i and j have been chosen (e.g. treatment A in row 1), k cannot in fact take any more than one value (column 3) and the correct number of plots, t^2, is thus accounted for. If we define R_j, C_k as the total of the yields of all plots in the jth row and kth column respectively in the layout, the sums of squares required are:

$$S_R = \frac{1}{t}\sum_{j=1}^{t} R_j^2 - \frac{G^2}{t^2}; \quad S_C = \frac{1}{t}\sum_{k=1}^{t} C_k^2 - \frac{G^2}{t^2}$$

S and S_T are similar to their previous forms, and since $r = t$, these become:

$$S_T = \frac{1}{t}\sum_{i=1}^{t} T_i^2 - \frac{G^2}{t^2}; \quad S = \sum_{i=1}^{t}\sum_{j=1}^{t}\sum_{k=1}^{t} y_{ijk}^2 - \frac{G^2}{t^2}$$

and finally S_E will be $S - S_T - S_R - S_C$. The degrees of freedom for S_T, S_R, S_C are each $(t-1)$; for S, (t^2-1), and so for S_E, $(t^2-1) - 3(t-1)$, reducing to $(t-1)(t-2)$. A numerical example of the Analysis of Variance, for the data of Example 16.4.1, is given in Table 16.4.

16.4.1 Example

Five different aptitude tests, A–E, are applied on five successive days to five different subjects who are considered comparable in intelligence. None of them has previously attempted tests of this type, and so it is required to remove any possible differences between days which could be attributed to a *learning* effect. This is done by using the layout of Fig. 16.2, and the scores obtained are listed in Table 16.4. Investigate the validity of the claim that the tests measure the same qualities, and also examine differences between subjects and between days.

In this layout, rows correspond to subjects and columns to days. The various sums of squares are: Tests, $S_T = \frac{1}{5}(331^2 + 332^2 + 393^2 + 310^2 + 304^2) - 1670^2/25$; Subjects, $S_R = \frac{1}{5}(318^2 + \cdots + 351^2) - 1670^2/25$; Days $S_C = \frac{1}{5}(321^2 + \cdots + 345^2) - 1670^2/25$; and $S = 56^2 + 62^2 + \cdots + 70^2 + 71^2 - 1670^2/25$. These values are shown in Table 16.5, and $S_E = S - S_T - S_R - S_C$ is obtained by subtraction as usual.

Table 16.4 Data of Example 16.4.1

Scores	Day					Subject totals	Totals for tests
	1	2	3	4	5		
Subject 1	E: 56	B: 62	A: 65	D: 59	C: 76	318	A: 331
2	C: 74	A: 65	D: 60	E: 61	B: 70	330	B: 332
3	B: 63	E: 59	C: 80	A: 66	D: 64	332	C: 393
4	A: 64	D: 63	B: 67	C: 81	E: 64	339	D: 310
5	D: 64	C: 82	E: 64	B: 70	A: 71	351	E: 304
Day totals	321	331	336	337	345	1670	1670

All F values are significant in this analysis. It is unlikely that any reasonable hypothesis could be set up about the way in which subjects differ, so it will be enough to record that they *do* appear to differ, and make no *t*-tests. As for days, there is a steady trend of increasing scores, which will cause no surprise, but the actual daily increases of score are not constant: there seems little point in *t*-tests here either, though perhaps a regression hypothesis (linear or logarithmic) would be worth a trial (as in Chapter 14). For tests A–E, a more detailed examination is now required. Significant differences are $t_{(12)}\sqrt{2\hat{\sigma}^2/5}$; the square root is $\sqrt{2 \times 1.97/5} = \sqrt{0.788} = 0.89$. The values of $t_{(12)}$ at the three standard significance levels are 2·179 (5%), 3·055 (1%) and

Table 16.5 Analysis of Variance for data of Example 16.4.1

Source of variation	D.F.	S.S.	M.S.	F-ratio
Subjects	4	118·0	29·50	$F_{(4, 12)} = 14·97$***
Days	4	62·4	15·60	$F_{(4, 12)} = 7·92$**
Tests	4	994·0	248·50	$F_{(4, 12)} = 126·14$***
Residual	12	23·6	1·97	
Total	24	1198·0		

4·318 (0·1%), and these values multiplied by 0·89 give the significant differences 1·94 (5%), 2·72 (1%) and 3·84 (0·1%). The mean scores for the various aptitude tests are: A, 66·2; B, 66·4; C, 78·6; D, 62·0; E, 60·8. Application of the significant differences to these means indicates that the tests form three groups: C, which gives the highest scores; A and B, considerably lower than C; D and E, which are somewhat lower than A, B. As usual, it is much more satisfactory to have definite hypotheses to examine, rather than making these comments on looking at the results; but as the three groups are all separate at the 0·1% level the danger of false conclusions is not very great. Nevertheless, if it had been thought that, say, A and E were much the same, some reflection on the result would be called for, and the structure of these two aptitude tests should then be re-examined.

For Latin Squares to be useful, they must not require an excessive number of replicates (remember that r must equal t); neither must f, the residual d.f., be too small. For $t = 4$ we have $f = 6$, and this may be just about adequate where the material being used is not too variable (see Section 16.6). Thus $t = 4$ and 8 are the lower and upper limits of common use of these designs: for $t = 3$, six replicates would be used, laid out as two 3×3 squares.

When choosing a layout for a single experiment, use should be made of Fisher and Yates'[2] tabulation of Latin Squares (Table XV) for $t = 4$, 5 or 6, while for larger sizes it may suffice to write a square down out of one's head and then allocate the treatments to letters at random.

16.5 Other designs

For some Latin Squares, it is possible to place a further set of treatments (usually denoted by greek letters—hence a *Graeco-Latin Square*) once in each row and column, and once with each of the treatments A, B, C, ... also. So three systematic sources of variation besides treatments can be taken out. These designs seem to be more readily applicable in industry than in biology, and the reader is referred to textbooks such as those by Cochran and Cox[10] or Clarke and Kempson[7].

In animal experiments, such as dietary trials, genetic differences among the animals (which would be the *unit-plots*) may be so large that any possible way of reducing the effect of these is very desirable. If blocks could be formed from animals of the same litter, all would be well; but very often these blocks prove to be much smaller than the required size of t units. Then designs known as *incomplete blocks*, in which not every treatment need appear in each block, are valuable, the *balanced* incomplete block being the simplest of these. Such designs lack orthogonality, and so are not easy to analyse. The reader is referred to standard textbooks for this topic also.

16.6 Precision of results

When we first considered significance tests, the usual three levels for testing were discussed, and reservations made about their uncritical use. We stressed the use of confidence intervals as an alternative. Nevertheless, too much faith is often placed in the very presence or absence of significance between a pair of means under test. This is equally true when comparisons are made between just two sets of data as in Chapter 8, or when the methods of experimental design described in Chapter 15 and earlier in this chapter are used. Two means may be declared significantly different if the numerical value of their difference exceeds $t_{(f)}\sqrt{2\sigma^2/r}$, the *significant difference* defined in Chapter 15; so that an experiment may be said to have good *precision* if $t_{(f)}\sqrt{2\sigma^2/r}$ is relatively small, and to have poor precision (to be *insensitive*) if this expression is large. An insensitive experiment is thus one in which only very large differences between means can be shown significant statistically, and is hardly likely to add to our knowledge about the treatments under study.

Precision in this sense can be improved by reducing the numerical values of some or all of the quantities in the expression $t_{(f)}\sqrt{2\sigma^2/r}$. The effect of *replication* is twofold. First, it improves precision to increase r, for the expression contains $1/\sqrt{r}$ and so can be, for example, halved by quadrupling r. Second, an increase in r helps to increase f, the residual d.f.; f also depends on whether the design is completely randomised, randomised blocks, or some other suitably chosen one. In its turn, f has two effects: one is to provide a well-based estimate of σ^2 (to which confidence-limits, if applied as in Chapter 11, would be reasonably narrow) and the other is to ensure that the values of $t_{(f)}$ to be used in significant differences are not too high. The latter is achieved if $f = 10$ or more, for a glance down the column headed $P = 0.05$ in the *t*-table (Table I) shows that the numerical value of $t_{(0.05,\ f)}$ increases quite rapidly as f decreases below 10 while it alters relatively little, and slowly, for f greater than 10. The amount of replication needed to give $f = 10$ or more depends on the type of design used and the number of treatments included; in theory, if there is a large number of treatments r need not be more than 2 (it must be at least 2, to give any replication on which to base σ^2), but in practice most workers would not be content with means based on less than four replicates.

However impractical, it might appear at first sight that 'the more the better' is the slogan for replication: but this is no wiser than most slogans, for the increase in precision obtained by raising f above 10 and r above 4 (or whatever its value must be so that f shall be 10) is often more than offset by an increase in σ^2. This can be due to several causes. It may no longer be possible to find a sufficient number of homogeneous units of material; a field experiment may have to go into two patches of land which have not had the same previous treatment, or different sources of seed may have to be used to sow the field. The experiment may become so unwieldy that inaccuracies creep into the recording, due either to fatigue or to pressure of time, or in some laboratory trials the last few samples examined may have deteriorated through being kept too long. Additional observers may have to take part in the recording, the chemical analysis of plant parts, the examination of microscope slides and so on, and this sometimes causes extra sources of systematic variation which, if they are not designed out, will inflate the value of σ^2.

The actual value of σ^2 is clearly important: it can be minimised by making sure that all systematic variation is allowed for (using, if needed, refinements of design like those mentioned in Section 16.5), by choosing plot sizes so that errors in recording are small in relation to the actual size of the measurement made (so that 10 blackcurrant bushes or 20 strawberry plants are commonly used as unit-plots in field experiments, giving a total crop per plot which is hardly affected by such recording errors as are reasonable), and also by a method known as the analysis of covariance, for which reference should be made to the text by Cochran and Cox[10] or Clarke and Kempson.[7] In greenhouse or laboratory work, conditions can sometimes be controlled sufficiently for σ^2 to be substantially smaller than could ever be expected in field experiments, and some relaxation of the lower limits to r and f may be permitted.

When σ^2 is known roughly, or can be guessed, from previous trials on the same or very similar crops, it is possible to say how many replicates will be needed to achieve a given degree of precision. Let us express the precision required by saying that if a pair of means differ numerically by an amount δ or more, we wish to show that they are statistically significantly different. This simply says that the significant difference $t_{(f)}\sqrt{2\sigma^2/r}$ must be less than (or equal to) δ.

16.6.1 Example

Suppose that σ^2 is approximately 25, and δ is required to be 6. Find the least value of r which will achieve this.

Unless r is known, f is not known exactly, so for a first approximation $t_{(f)}$ is replaced by 2. Then $2\sqrt{50/r} \leqslant 6$, so that $\sqrt{r} \leqslant \frac{2}{6}\sqrt{50}$ or $r \geqslant 50/9$, and since r has to be a whole number this gives $r = 6$ as the minimum replication needed to achieve the stated aim. It is possible that when the design has been chosen and the number of treatments settled, so that f and $t_{(f)}$ can be quoted exactly, r may need increasing to the next whole number. But the stated aims are only approximate, and we do not in any case know whether the estimate of σ^2 from this particular experiment will be very near to 25, since the value 25 can only be inferred from previous experience with the same type of experimental unit and the variance is bound to change from one experiment to another. This method should be regarded as giving an idea of what order of replication is needed—whether r must be large, say 10 or 12, or a more reasonable value such as 4 or 5. It may lead to an inconveniently large value of r, and so to a reconsideration of whether the stated aim for precision is necessary. And it may lead also to a decision that an experiment with the resources available is just not worth doing for the value of information likely to come from it—one suspects that this decision is not taken nearly as often as it should be.

16.6.2 Example

Sometimes we are interested in detecting whether a treatment changes a mean *by some stated percentage*: for example, can we increase the yield of a crop by *at least* 7% by adding an extra fertiliser element? The idea here is that unless we can achieve at least a 7% increase it will not pay the cost of the extra fertiliser. So the requirement of precision in this situation may be expressed as the ratio

$$\frac{\text{increase in crop yield by using new treatment}}{\text{average crop yield under standard treatment}}$$

In order to carry out the analysis in this type of problem we need not simply an estimate (or guess) of σ^2, but instead the ratio of σ to the mean yield of the standard treatment. This ratio σ/mean is called the *coefficient of variation*, V. Suppose it is 16% in our example: that is to say we are told $\delta/\text{mean} = 7\%$ and $\sigma/\text{mean} = 16\%$. Thus $t_{(r)}\sigma\sqrt{2/r} \leqslant \delta$ gives $t_{(r)}\sqrt{2/r} \times 16 \leqslant 7$, or approximately $32\sqrt{2} \leqslant 7\sqrt{r}$, giving $r = 42$ approximately. So we are asking for too high a level of precision to be reasonable, with material of this degree of variability.

It is not unusual in field experiments for the coefficient of variation, V, to reach 20%, but in laboratory and greenhouse studies something like $V = 5$ to 10% should usually be attainable, so that to achieve precision equivalent to detecting a difference $|\delta| = 10\%$ should be reasonable in the latter case: a calculation like the previous one suggests a quite practicable value for r. Using this r, we would expect that if the mean yield under the standard treatment was 100, and a new treatment could raise this to at least 110, the difference between these two treatments would usually show up as significant.

16.7 Randomised block analysis in MINITAB

TWOWAY analysis (treatments and blocks as sources of systematic variation) takes a set of data in the form of Table 16.2, where we must number the rows (blocks) 1, 2, . . . and the columns (treatments) 1, 2, . . . to identify each item of data by its suffices i, j. Among various other forms of output we can obtain the Analysis of Variance of Table 16.1, in which the two sources of variation can be assigned names (blocks and treatments, or for example observer and suspension as in Example 16.2.1), and Residual is called *Error*. Mean squares are provided and we may continue our own analysis from there.

It is possible to output the *fitted values* predicted on the model $y = \mu + t_i + b_j$, and the residuals e_{ij} on each plot which are the differences between observed and fitted values. These can be very useful in drawing attention to any plots that were unusual, and further examination of the block and/or the treatment containing a plot with a large residual can be helpful (see Clarke and Kempson[7]).

16.8 Exercises on Chapter 16

16.8.1 Four different plant densities, A–D, are included in an experiment on the growth of lettuce. The experiment is laid out as a randomised block, and the same number of plants is harvested from each plot, giving the weights (kg) recorded below. Examine whether density appears to affect yield. (It may be assumed that planting density increases in the order A, B, C, D.)

Density	Block					
	I	II	III	IV	V	VI
A	2·7	2·6	3·1	3·0	2·5	3·0
B	3·0	2·8	3·1	3·2	2·8	3·1
C	3·3	3·3	3·5	3·4	3·0	3·2
D	3·2	3·0	3·3	3·2	3·0	3·1

16.8.2 Six chemical compounds A–F are being tested as potential protective fungicides against apple mildew. Thirty plants of the same variety of apple seedling are available; these are split into six groups of five, and each group of five is then dipped into a suspension of one of the compounds A–F. Afterwards the whole collection of plants is placed in a closed environment which is infected with mildew spores. The plants are positioned in five randomised blocks to take out any effects due to slight systematic variations in the environment, and subsequently the number of lesions on the first leaf on each plant is counted. These records are tabulated below. Examine the differences between compounds.

	Block				
Compound	I	II	III	IV	V
A	32	22	26	25	18
B	5	8	6	2	4
C	19	12	15	11	14
D	11	3	7	10	8
E	26	16	22	20	13
F	4	3	8	5	1

16.8.3 A supermarket organisation buys in a particular foodstuff from four suppliers A, B, C, D and subjects samples of this to regular tasting tests by expert panels. Various characteristics are scored, and the total score for the product is recorded. Four tasters *a*, *b*, *c*, *d* at four sessions obtained the results below. Analyse and comment on these.

	Taster			
Session	*a*	*b*	*c*	*d*
1	A: 21	B: 17	C: 18	D: 20
2	B: 20	D: 22	A: 23	C: 19
3	C: 20	A: 24	D: 22	B: 19
4	D: 22	C: 21	B: 22	A: 26

16.8.4 The mean estimated volumes of wood growth (m^3) in their first season were measured for a *Pinus* species planted at four different densities A–D. The experiment was laid out as five randomised blocks. The data are given below.

Is there evidence that blocking was necessary? What implication does this have for further similar experiments?

The densities increased from A up to D. Test whether there are differences between the means of the four treatments and suggest any further analysis that might be useful.

	Block				
Density	I	II	III	IV	V
A	4·2	5·0	3·7	4·4	4·6
B	4·8	5·2	4·5	4·9	4·7
C	5·2	4·8	6·3	5·5	5·7
D	5·4	5·9	6·0	6·2	6·5

16.8.5   The mean height to which wheat plants grow in a fixed period of time is 18 cm, and the standard deviation of height is 2·8 cm. An experiment using eight treatments is laid out in four randomised blocks. What is the least difference in mean height between two treatments that can be shown as significant in this experiment?

16.8.6   In a field experiment using plots of 4 m^2, it is known from past experience that the coefficient of variation of crop yield will be about 8%. What is the minimum replication that will enable a 10% difference in treatment mean yields to be detected in a significance test?

   If plots of 2 m^2 are used instead, so that this replication can be doubled, the coefficient of variation will go up to 12%. What difference can now be detected as significant?

17 Simple factorial experiments

17.1 Factors and interaction

So far we have not considered that the treatments in an experiment may be related to one another; all the designs described would, for example, suit perfectly well for comparing a variety of fungicides which are chemically quite different in structure, and have only one similarity, namely that they all act against the same fungus upon which the experiment is being conducted. But it is often necessary to make up a set of treatments from basic *factors*, for example in fertiliser trials. A fertiliser needs to contain the elements N, P, K, Mg (nitrogen, phosphorus, potassium, magnesium), and the treatments examined in an experiment will be various combinations of amounts of these elements. However, the real topic of interest is not these particular combinations but the basic factors, the elements themselves: the most interesting question of all is whether the factors act independently of one another. If we consider how much N to apply, and then how much P, it is very likely that the best amount of P will depend very closely on how much N is present (whereas if fungicides are included in the same experiment, to keep plants *clean*, it is much less likely that the N or P requirement of a plant would be affected by which fungicide had been used on it).

As an illustration of the fundamental idea of **interaction**, suppose that a fertiliser containing only basic amounts of N and P induces an average yield of five units in a crop; and that this can be increased to 30 units or 20 units respectively by adding either extra N or extra P respectively, *without* the other. The response to extra N is thus 25 units (30 minus 5) and to extra P is 15 units (20 minus 5). But this gives an incomplete picture until we know what happens when the extra amounts of N and P are both added *together*. If N and P act independently, the response to N and P together will be the sum of the two responses already measured, namely 25 plus 15, i.e. 40 units, so that the actual yield (basic plus response) is 45 units. However, we are very likely to find that the response to N can be enhanced by adding P as well, so that with both together the yield proves to be *more* than 45 units: we then say that N and P *interact*, and in this case the interaction is a positive one. Some combinations of elements can result in a depression of the predicted yield, which would be called a negative interaction, when they are present in quantities that give the plants an unbalanced nutrient regime: the elements K and Mg in fertilisers can give this effect, and even be toxic when the unbalance is severe.

Whenever it seems likely that two or more basic elements or factors will interact, little or no useful information is obtained by experimenting on them one at a time. They must be examined together, in a design known as a *factorial experiment*, which provides information on how the factors interact, either in pairs (NP, NK, etc.) or larger groups (NPK, etc., wherein the extent of interaction between N and P can be altered by altering the amount of K used, and so on).

Notation

There is a standard notation for factorial experiments, which unfortunately clashes to some extent with that traditionally used for experiments in general, as described in earlier chapters. In factorial experiments the factors are denoted by capital letters A, B, C, . . .; the actual treatments used in the conduct of a factorial experiment will consist of combinations of various amounts (*levels*, as they are called) of these factors, and we can no longer denote the treatments by capitals.

17.2 Experiments with two-level factors

A useful type of factorial experiment, which can be employed at the outset of a programme of work to discover just which out of a whole set of possible factors do interact with one another, has each factor included at two levels only: high and low, or present and absent. If, for example, there are just three likely factors to be investigated, we shall call them A, B and C; and the first experiment in a series to study their behaviour will contain a high and low level of each. These levels will be used in all possible combinations, of which there are $2 \times 2 \times 2$; there is a standard method of denoting these, as follows. Let the symbol *abc* stand for that combination which has A, B, C all present (or all at the high level); *ab* has A, B present and C absent (or A, B at the high level and C at low); *bc* has B, C present and A absent; *ac* has A, C present and B absent; *a* has A present, B and C absent; *b* has B present, A and C absent; *c* has C present, A and B absent; ① has all three absent.* These $2 \times 2 \times 2$, or 2^3, that is eight, treatment combinations will then be laid out in an experimental design appropriate to the conditions under which the work is to be done (e.g. in randomised blocks, replicated a reasonable number of times). It is only when we come to consider the seven degrees of freedom for treatments that we alter the construction of the analysis of variance so as to show the effects of, and interactions between, factors. These 7 d.f. correspond to the sum of squares between the totals of the eight treatment combinations ①, *a*, *b*, *c*, *ab*, *ac*, *bc*, *abc*. It is possible to define seven single degrees of freedom: three for **main effects**, denoted by the capital letters A, B, C; three for the **first-order interactions** between pairs of elements, denoted by AB, AC, BC; and one for the **second-order interaction** of all three elements, ABC. Note here again that the capital letters used to denote the factors in an experiment are also used to denote the main effects, but *not* the actual combinations of factor levels that make up the treatments applied to the experimental plots: these are always written in small letters.

The main effect of A is the average response to addition of A, averaged over all the combinations of other factors (B, C) in the experiment. It is thus $\frac{1}{4}[(abc - bc) + (ab - b) + (ac - c) + (a - ①)]$, since each bracket ( ) represents a response to A, i.e. a difference between a plot which contains A and one which does not, both having the same levels of B and C; there are four such brackets to be averaged. Obviously this is equivalent to taking the difference between all plots with A and all without, so that

---

* The symbol ① here denotes what is usually written (1).

$$A = \tfrac{1}{4}[(abc + ab + ac + a) - (bc + b + c + ①)], \quad \text{and similarly}$$
$$B = \tfrac{1}{4}[(abc + ab + bc + b) - (ac + a + c + ①)],$$
$$C = \tfrac{1}{4}[(abc + ac + bc + c) - (ab + a + b + ①)].$$

The interaction AB is defined as the average response to A in the presence of B minus the average response to A in the absence of B, averages in each case being taken over all possible combinations of the remaining factors in the experiment (in this example, just the presence and absence of C). Thus we obtain

$$AB = \tfrac{1}{4}[\{(abc - bc) + (ab - b)\} - \{(ac - c) + (a - ①)\}]$$

i.e.

$$AB = \tfrac{1}{4}[(abc + ab + c + ①) - (bc + b + ac + a)].$$

AC and BC may be obtained by similar arguments, but the expressions for interactions are not easy to remember, and fortunately there exists a convenient algebraic way of finding out what expression to use for any main effect or interaction of any order; it gives the following results.

Main effects:	$A = \tfrac{1}{4}(a-1)(b+1)(c+1)$
	$B = \tfrac{1}{4}(a+1)(b-1)(c+1)$
	$C = \tfrac{1}{4}(a+1)(b+1)(c-1)$
First-order interactions:	$AB = \tfrac{1}{4}(a-1)(b-1)(c+1)$
	$AC = \tfrac{1}{4}(a-1)(b+1)(c-1)$
	$BC = \tfrac{1}{4}(a+1)(b-1)(c-1)$
Second-order interaction:	$ABC = \tfrac{1}{4}(a-1)(b-1)(c-1)$

Write down the brackets $(a\ 1)$ etc. for every factor used in the experiment, and whenever a capital letter appears in the effect or interaction being calculated, put a minus sign into the corresponding bracket. Plus signs go into the other brackets and have the effect of averaging over both levels of the factors represented by these brackets. The expressions are evaluated just as though they were simple algebraic products, e.g.

$$AB = \tfrac{1}{4}(a-1)(bc + b - c - 1) = \tfrac{1}{4}(abc + ab - ac - a - bc - b + c + ①)$$

as in the definition. Although verbal definitions for higher-order interactions are complicated, these too can be calculated in this way, and we obtain

$$ABC = \tfrac{1}{4}(abc - ab - ac - bc + a + b + c - ①).$$

Whenever interactions are considered possible, factorial designs must be used; but if interactions are not found, nothing is lost, for the main effects still represent the results of a well-replicated trial on each element separately, comparing, for A, just the two levels *present* and *absent*, each level based on half the total number of plots in the whole experiment.

After preliminary 2^n experiments (*n* factors each at two levels), those factors that prove of interest will need much closer study, and three-level experiments, which are common, or higher numbers of levels of a few factors, are needed. We give a simple example of such an experiment later in this chapter, but the general theory of these, and especially how to proceed when the number of treatment combinations is very large, is not easy and readers should consult a more advanced text such as those by Cochran and Cox[10] or Clarke and Kempson.[7]

17.2.1 Example. A 2^3 experiment

Table 17.1 summarises the increases in weight of guinea-pigs fed on basic or supplemented diets; three different supplementing factors, bran, vitamins and antibiotics, were being examined. Each diet was fed to four animals, one in each of four pens; thus the experimental layout was randomised blocks, with pens forming blocks.

Table 17.1

Diet	Basic	+ Anti-biotics (A)	+ Bran (B)	+ Vita-mins (V)	+ A and B	+ A and V	+ B and V	+ A, B and V	Pen totals
	①	a	b	v	ab	av	bv	abv	
Pen I	7	8	11	10	12	12	15	16	91
II	8	10	11	11	15	16	16	15	102
III	6	8	13	12	14	12	13	15	93
IV	7	8	12	10	14	14	17	16	98
	28	34	47	43	55	54	61	62	384

Table 17.2 gives the analysis of the layout as it was, in randomised blocks, ignoring the factorial nature of the treatments: from it we obtain an estimate of σ^2 in the usual way (from the residual error-mean-square) and also a test of whether segregation into

Table 17.2 First-stage Analysis of Variance

Source of variation	D.F.	S.S	M.S.	
Pens	3	9·250	3·0833	$F_{(3, 21)} = 2\cdot42$ n.s.
Treatments (Diets)	7	268·000	38·2857	
Residual	21	26·750	1·2738	
Total	31	304·000		

pens removed any significant amount of variation, which in the example it did not. Seven single d.f. are now obtained; in the definition of any particular effect or interaction there are four treatments carrying a plus sign and four a minus, e.g. for A, each of abv, ab, av, a appears with a plus and b, v, bv, ① with a minus. The total for A is thus $62 + 55 + 54 + 34 - 47 - 43 - 61 - 28 = +205 - 179 = +26$, and the single d.f. for A has value $(+26)^2/32$, 32 being the complete number of experimental plots (animals). Similarly the total for B is $+225 - 159 = +66$, for V $+220 - 164 = +56$, for AB $+188 - 196 = -8$, for AV $+191 - 193 = -2$, for BV $+185 - 199 = -14$, and for ABV $+186 - 198 = -12$. Squaring each of these and dividing by 32, we obtain the single d.f. listed in Table 17.3, and as a check they do indeed add to the seven d.f. *treatments* term already obtained. In this method, the factor $\frac{1}{4}$ in the original definition is not used (though of course it is included in the divisor 32).

We now see that each of the main-effect terms has a very large value of F, significant at 0·1%; but on looking at the rest of the table we discover an interaction, BV, which is

Table 17.3 Full Analysis of Variance

Source of variation	D.F.	S. S.	M.S.	
Pens	3	9·250	3·0833	$F_{(3, 21)} = 2·42$ n.s.
A	1	21·125	21·1250	$F_{(1, 21)} = 16·58$***
B	1	136·125	136·1250	$F_{(1, 21)} = 106·86$***
V	1	98·000	98·0000	$F_{(1, 21)} = 76·93$***
AB	1	2·000	2·0000	$F_{(1, 21)} = 1·57$ n.s.
AV	1	0·125	0·1250	$F < 1$ n.s.
BV	1	6·125	6·1250	$F_{(1, 21)} = 4·81$*
ABV	1	4·500	4·5000	$F_{(1, 21)} = 3·53$ n.s.
Treatments	7	268·000		
Residual	21	26·750	1·2738	
Total	31	304·000		

significant. Thus, while it is correct to examine the response to A in terms of its main effect the responses to B and V have to be considered together, and it would be misleading to say anything specific about their main effects. If we consider A first, the total weight increase of the 16 animals receiving A was 205, so that the mean per animal was 12·8; the mean increase without A was similarly 179/16 = 11·2, and so the response to A is a beneficial one. For B and V, we need a *two-way* table of means: eight animals received neither B nor V, their total weight increase being 62 (*a* and ①), and the mean thus 7·75; the mean with B but not with V (*b* and *ab*) is 12·75, with V but not B (*v* and *av*) 12·13, and with both B and V (*bv* and *abv*) 15·38. The significant differences between any two of these means based on eight observations are $t_{(21)}\sqrt{2\hat{\sigma}^2/8} = t_{(21)}\sqrt{\frac{1}{4} \times 1·2738} = 0·564 \times t_{(21)}$. Putting in the tabular values of $t_{(21)}$, namely 2·080 (5%), 2·831 (1%), 3·819 (0·1%), we obtain the significant differences 1·17 (5%), 1·60 (1%) and 2·15 (0·1%). A table of means, Table 17.4, is now written out. The B-effect is significant at 0·1% both in the presence of V and in its absence, as is seen by looking at the two rows of the table in turn; similarly, the V-effect is significant at 0·1%

Table 17.4 Two-way table of means for B and V

	−B	+B
−V	7·75	12·75
+V	12·13	15·38

for either −B or +B, and the only reason for an interaction proving significant here seems to be that the sizes of the B- and V-effects, although always very significant, vary somewhat numerically.

When a two-level factorial experiment has indicated which single factors or groups of factors have a significant effect on the records that have been analysed, a more detailed study of the form of the effect is required. In Example 17.1, B and V must be examined together in a further experiment wherein each factor is present at three levels at least, though since A did not appear to affect either of them there would be no need to study it at the same time.

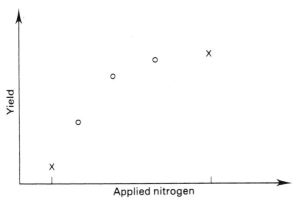

Fig. 17.1 Effect of amount of nitrogen fertiliser on crop yield; two points x obtained in a preliminary experiment, and three points ○ in later detailed study.

A single factor (such as A above) showing a main effect in a two-level experiment requires to be studied further at several levels, with a view to establishing a definite equation relating its level to the response y. As an example, consider the amount of nitrogen present in a fertiliser applied to plants; suppose that the two points marked x in Fig. 17.1 represent the levels used in the original experiment. There are many forms which a curve relating yield y to applied level N might take between these two points. A reliable curve can hardly be based on less than five observations, so that in further experiments at least the three points marked ○ should be added, giving a minimum of five levels of N studied simultaneously. Curves similar to that in Fig. 17.1 are common in studies of plant growth; on occasion also something very near to a straight line may be obtained, or even a curve which has a maximum yield at an intermediate level of N and begins to fall again at the higher levels investigated.

If two factors, say N and P in a fertiliser, interact, we expect to find a different form of response (crop yield, etc.) to P according to whether N has a higher or lower level. Let us suppose that N is present at three levels and P at five; then Fig. 17.2 shows the type of response curve which may arise. When more than two factors are involved, it

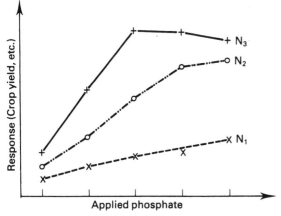

Fig. 17.2 Typical responses to phosphate fertiliser when nitrogen is present at three different levels N_1, N_2, N_3.

becomes progressively more difficult to interpret the situation from diagrams, and it is necessary to fit an equation to the data. The examination of growth curves in general in biology is an extensive subject, and some of the methods of proceeding are explained by Cochran and Cox.[10]

17.3 Mixed factorials

When factors are not all used at the same number of levels, we call the experimental design a *mixed factorial*. We give below the analysis for a design using three fertilisers N, P and K on strawberries, in which N was included at two levels and each of the other factors at three levels. This is a '2 × 3 × 3' experiment, because there are 2 × 3 × 3 = 18 different combinations of levels of N, P, K that we can think of; these 18 combinations will form the treatments in the experiment. Factors at more than two levels need a more general notation than we have so far had to use: we shall call the three levels of P P_1, P_2, P_3 and those of K K_1, K_2, K_3. To avoid confusion we shall write the two levels of N in the same notation as N_1, N_2.

17.3.1 Example
The amount of nitrogen (ppm) present in dried leaves of strawberries was measured from an experiment having two levels of N and three each of P and K, the full set of treatments being replicated twice in randomised blocks. Table 17.5 gives the list of treatment combinations and the detailed records of the leaf analysis. Treatment

Table 17.5 Leaf nitrogen (ppm) present in dried strawberry leaves from 2 × 3 × 3 N × P × K fertiliser trial

Treatment	Block I	Block II	Total
$N_1 P_1 K_1$	2·42	2·45	4·87
$N_1 P_1 K_2$	2·46	2·44	4·90
$N_1 P_1 K_3$	2·48	2·51	4·99
$N_1 P_2 K_1$	2·43	2·44	4·87
$N_1 P_2 K_2$	2·48	2·46	4·94
$N_1 P_2 K_3$	2·52	2·51	5·03
$N_1 P_3 K_1$	2·47	2·50	4·97
$N_1 P_3 K_2$	2·50	2·53	5·03
$N_1 P_3 K_3$	2·54	2·55	5·09
		Total for N_1:	44·69

Treatment	Block I	Block II	Total
$N_2 P_1 K_1$	2·48	2·50	4·98
$N_2 P_1 K_2$	2·47	2·52	4·99
$N_2 P_1 K_3$	2·53	2·55	5·08
$N_2 P_2 K_1$	2·47	2·51	4·98
$N_2 P_2 K_2$	2·52	2·50	5·02
$N_2 P_2 K_3$	2·55	2·52	5·07
$N_2 P_3 K_1$	2·50	2·54	5·04
$N_2 P_3 K_2$	2·56	2·51	5·07
$N_2 P_3 K_3$	2·57	2·55	5·12
		Total for N_2:	45·35

Total for Block I = 44·95, for Block II = 45·09

Grand Total G = 90·04

$N_1P_1K_1$ denotes the lowest level of each factor applied to the plants, $N_2P_3K_3$ the highest level of each.

In the Analysis of Variance, we expect, as before, to be able to set up main-effect terms for each of the factors N, P, K and also terms for interactions between any pair of these and between all three. There are only two levels of N, and so the main effect of N will have one degree of freedom as before. But P and K have each been included at three levels; so for each of these factors the main effect must measure the difference between three sets of plots—those receiving level 1, those receiving 2 and those receiving 3. We noted in Chapter 15 that if there are t treatments being compared, $(t-1)$ d.f. will be used in making the comparison. In the same way, in a factorial experiment, the d.f. for the main effect of any factor will be one less than the number of levels of that factor. P and K in this example both have two d.f. The rule for interactions is that d.f. for NP = (d.f. for N) × (d.f. for P), and similarly for NK, PK; and in the same way d.f. for NPK = (d.f. for N) × (d.f. for P) × (d.f. for K). Therefore in this example NP and NK each have two d.f., while PK and NPK each have four d.f. Now that we have assigned degrees of freedom to all the possible main effects and interactions of the three factors, we add these d.f. together and check that they do account for all the 17 d.f. between the 18 treatment combinations used in the experiment.

In order to calculate main effects we shall need the totals level-by-level; for interactions we shall need two-way tables. So we extract from Table 17.5 three two-way tables, which are shown in Table 17.6. In the table for N and P, the entry in the

Table 17.6 Two-way tables showing totals summed over all blocks and all levels of the third factor

	P_1	P_2	P_3		N totals
N_1	14·76	14·84	15·09	:	44·69
N_2	15·05	15·07	15·23	:	45·35
P totals	29·81	29·91	30·32		90·04 = G

	K_1	K_2	K_3		N totals
N_1	14·71	14·87	15·11	:	44·69
N_2	15·00	15·08	15·27	:	45·35
K totals	29·71	29·95	30·38		90·04 = G

	K_1	K_2	K_3		P totals
P_1	9·85	9·89	10·07	:	29·81
P_2	9·85	9·96	10·10	:	29·91
P_3	10·01	10·10	10·21	:	30·32
K totals	29·71	29·95	30·38		90·04 = G

first row and first column is the sum of *all* plots that had the lower level 1 of N and the lowest level 1 of P, from both blocks and all three levels of K: it is thus the sum of six plot measurements. The other five entries in the table are made up in the same way. In the margins of the table are shown the overall totals for N_1 and N_2 (rows) and also for P_1, P_2 and P_3 (columns). The NK table is made up in just the same way; so is the PK table, except that each entry here is the sum of four plot measurements (from both blocks and two levels of N).

The analogy between d.f. for t treatments and d.f. for a main effect gives us the key to computing the sum of squares for a main effect: we simply use the totals for each level and calculate the sum of squares between them. Thus for P, the totals are 29·81 (P_1), 29·91 (P_2) and 30·32 (P_3), which add to $G = 90·04$. N, the total number of plots, is 36, and each P total is based on 12 plots. The sum of squares for the main effect of P is thus $\frac{1}{12}(29·81^2 + 29·91^2 + 30·32^2) - 90·04^2/36$. That for K is $\frac{1}{12}(29·71^2 + 29·95^2 + 30·38^2) - 90·04^2/36$ and, for N, $\frac{1}{18}(44·69^2 + 45·35^2) - 90·04^2/36$, since each N total is based on 18 plots. The values of all these are shown in the Analysis of Variance, Table 17.7.

Turning to interactions between two factors, we ask what the NP two-way table of totals, in Table 17.6, can tell us besides the information about main effects of N and P that we have already extracted from the margins. The situation is exactly like that in a 2×2 table: if the entries in the body of the table can be predicted as soon as we know the margins, then the factors do not interact, but if (for example) the pattern of P response under N_1 is very different from that under N_2 we need to be given all the

Table 17.7 Analysis of Variance for Example 17.3.1

Source of variation	D.F.	S.S.	M.S.	
Blocks	1	0·000 54	0·00 054	$F_{(1,17)} = 1·29$ n.s.
N	1	0·01 210	0·01 210	$F_{(1,17)} = 28·81$***
P	2	0·01 217	0·00 609	$F_{(2,17)} = 14·50$***
K	2	0·01 921	0·00 961	$F_{(2,17)} = 22·88$***
NP	2	0·00 095	0·00 048	$F_{(2,17)} = 1·14$ n.s.
NK	2	0·00 071	0·00 036	$F_{(2,17)} < 1$ n.s.
PK	4	0·00 053	0·00 013	$F_{(4,17)} < 1$ n.s.
NPK	4	0·00 019	0·00 005	$F_{(4,17)} < 1$ n.s.
Treatments	17	0·04 586		
Error	17	0·00 716	0·00 042	
Total	35	0·05 356		

separate NP totals—the margins are not enough—and there *is* an interaction between N and P. The six NP totals here contain the information on the N main effect (when added by rows), the P main effect (when added by columns) and, in addition, the NP interaction. There are six totals, so five d.f. for comparing them; one d.f. is the N main effect, two d.f. are the P main effect, and the remaining two d.f. are the NP interaction. The five d.f. sum of squares from these six totals is $\frac{1}{6}(14·76^2 + 15·05^2 + 14·84^2 + 15·07^2 + 15·09^2 + 15·23^2) - 90·04^2/36 = 0·025\ 22$, each total being based on six observations; this is $N + P + NP$, from which we easily compute NP. Similar calculations from the other two tables in Table 17.6 give NK and PK, whose values are shown in Table 17.7.

It now only remains to calculate the three-factor interaction NPK. The best way to do this is to work out the total sum of squares for the whole set of 18 treatment combinations used in the experiment, and to subtract from that the sums of squares for each main effect and two-factor interaction already found. The 18 treatment totals (each one the sum of just two entries, from Block I and Block II) are shown in Table 17.5, and give the sum of squares $\frac{1}{2}(4·87^2 + 4·90^2 + \cdots + 5·07^2 + 5·12^2) - 90·04^2/36 = 0·045\ 86$. The Analysis of Variance is completed by working out the total sum of squares for all 36 plots, $(2 \cdot 42^2 + 2 \cdot 45^2 + \cdots + 2 \cdot 57^2 + 2 \cdot 55^2) - 90 \cdot 04^2/36$ and the sum of squares for blocks.

From Table 17.7, we see that there is no evidence of any interaction, the only significant terms in the analysis being the three main effects. At once it is clear that N_2 gives higher results than N_1, since there are only two N levels to compare. But there are three P levels, so we require the three means: P_1 has mean $29 \cdot 81/12 = 2 \cdot 484$, P_2 has mean $2 \cdot 493$ and P_3 has mean $2 \cdot 527$. Each of these means is based on 12 observations, and so we may compare them by working out the least significant difference $t_{(17)} \sqrt{2\hat{\sigma}^2/12} = t_{(17)} \sqrt{0 \cdot 000 42/6} = 0 \cdot 008 37\, t_{(17)}$. The values of $t_{(17)}$ from Table I are $2 \cdot 110$ (5%), $2 \cdot 898$ (1%), $3 \cdot 965$ (0·1%) and so significant differences are $0 \cdot 018$ (5%), $0 \cdot 024$ (1%) and $0 \cdot 033$ (0·1%). From these we see that P_1 and P_2 do not differ significantly in mean, but P_3 is greater than either of these at the 0·1% significance level. The means for K are also based on 12 observations each, and so we can use the same significant differences; K_1 mean $= 2 \cdot 476$, K_2 mean $= 2 \cdot 496$, K_3 mean $= 2 \cdot 532$. This time K_1 is below K_2 at the 5% significance level, and K_3 is greater than either at the 0·1% significance level. This completes the summary of results for this Example, but if there had been any significant interactions in the Analysis of Variance we would have dealt with them as in Example 17.2.1 (see Table 17.4 and the explanation of how to test the means in that two-way table).

17.4 MINITAB with two factors

With a Randomised Block routine there is scope for analysing two main effects equivalent to blocks and treatments. By specifying *levels* for these classifications we can turn it in to a program for two factors with interaction, and the output will contain rows for Factor A, Factor B, Interaction AB and Residual (Error). The two factors can each be at two or more levels, but single degrees of freedom are *not* calculated in the Analysis of Variance. A two-way table of means for all the combinations of levels of the two factors is given, similar to Table 17.4 but with as many rows and columns as there are levels. From this a graph like Fig. 17.2 can be drawn. The interaction term may be omitted by including a subcommand ADDITIVE, which then fits a model like a randomised block.

17.5 Exercises on Chapter 17

17.5.1 A fertiliser trial on strawberries consists of four replicates of the four treatments ①, *n*, *p*, *np*, combinations of high and low levels of nitrogen and phosphorus. The resulting crop yields per plot (in suitable units) are:

Block	①	*n*	*p*	*np*
I	13	24	16	27
II	12	25	14	34
III	18	24	15	32
IV	15	31	20	30

Carry out an analysis to determine the effects (if any) due to blocks, N, P and the interaction NP.

17.5.2 Two spray materials A, B, are applied to apple trees; the effect of A is to reduce damage to fruit due to scab, and B controls mildew on shoots but may damage the skin of the fruit. Four treatments (A alone, B alone, both together, neither) are employed in five replicates, and the appearance of a sample of fruit is assessed on a scoring scale with the following results (low scores imply poor appearance):

Block	Neither A nor B	A alone	B alone	Both A and B
I	12	24	10	12
II	15	28	12	15
III	8	23	10	19
IV	17	20	14	11
V	13	30	9	16

Analyse and comment on these results.

17.5.3 Seedling plants are transplanted from a greenhouse to an outdoor plantation at three different times, T_1, T_2 and T_3, being given one of four different hormone rooting treatments R_1, R_2, R_3, R_4 when transplanted. Two plants are used for each RT combination, the layout being completely randomised. The results are summarised in the following table: the entry 11; 13 for the $R_1 T_1$ combination indicates that one plant grew 11 cm and the other 13 cm in a fixed period of time after transplanting; and so on.

	T_1	T_2	T_3
R_1	11;13	18;12	18;17
R_2	15;21	20;23	15;16
R_3	18;18	21;20	19;20
R_4	19;16	19;20	26;24

Analyse and comment on these results.

17.5.4 In a fruit storage experiment, three varieties of apple were kept for four different lengths of time at each of two temperatures. The percentage showing some deterioration was estimated in two samples as each group was removed from store. The percentages for the two samples in each combination are shown as '10, 12' etc. in the following table.

Temperature	lower			higher		
Variety	B	C	L	B	C	L
Time 1	10,12	6, 9	14,16	15,19	8,11	15,20
(months) 2	14,23	10,17	19,25	19,23	14,17	22,24
3	20,26	18,21	23,27	27,32	19,24	30,33
4	28,33	25,29	30,35	35,36	27,31	34,37

Examine the effects and interactions of the different factors, and illustrate the results for each temperature graphically.

18 Missing observations and non-normality

18.1 Missing observations

When an experiment has been designed according to statistical principles, it may be analysed by one of the relatively simple methods already described. But these methods are based on totals and means of groups of observations; thus they cannot be applied when some of the plots in the layout cannot be recorded. If this happens, the totals are incomplete, in the sense that (for instance) not every treatment is represented in every block total if the layout was randomised blocks. At the end of Chapter 16 it was pointed out that incomplete block designs are much less easy to analyse than the orthogonal ones considered in this book. Fortunately, as we shall see below, it is fairly easy to deal with the difficulties which arise when only one or two plots in an orthogonal layout cannot be recorded: these we describe as *missing plots* or *missing observations*. If one of the treatments in a randomised block fails completely it may simply be omitted from the analysis, since it affects every block equally; and if one whole block is lost, similar omission is made.

Often, one plot may be lost for a reason not attributable to treatment: plants in a field or greenhouse experiment on fertilisers may become diseased; there may be accidental damage by cultivation machinery; an analytical sample or a microscope slide may be lost or damaged in a laboratory; an animal in a nutrition trial may become ill through disease. Before we treat a lost observation as a 'missing value' we must be quite sure that the loss has not been caused by the experimental treatment: if in a fertiliser experiment plants die on plots which have received a fertiliser that lacked some essential nutrient element, the value 0 is the correct record of crop yield; in this case much useful information might be obtained by recording features other than yield, such as the heights to which plants have grown, numbers of leaves formed, and so on—records which would *not* read 0 for these plots and would help to pinpoint the effect of fertiliser deficiency.

We will assume in what follows that only one plot has been lost and that it is clear beyond reasonable doubt that the loss has been accidental or independent of treatment. The solution in this case follows these three steps:

1 a value is 'estimated' for the observation on the lost plot;
2 this estimate is written into the table of observed records and the analysis now carried out by exactly the same method as usual, i.e. as described in Chapter 16, for the design which has been used; *except* that
3 the degrees of freedom for total and residual sums of squares are each reduced by 1.

The required estimate of the missing observation depends on the type of design used.

18.2 Randomised blocks and Latin Squares

The table of observations is written out as in Example 16.2.1, and has one blank space where the observation has been lost; this will affect one particular block and one particular treatment, whose totals are incomplete because they are calculated from one observation fewer than they should be. Call these incomplete totals B' and T' respectively; the grand total is of course similarly incomplete, and we label it G'. The estimate, x, of the lost observation uses just these incomplete totals, as follows:

$$x = \frac{rB' + tT' - G'}{(r-1)(t-1)}$$

This value of x is written into the blank space, totals recalculated, and the analysis continued as normal except that the total sum of squares has $(rt-2)$ rather than $(rt-1)$ d.f., and the residual d.f. are also reduced by 1.

18.2.1 Example
Four diets, A–D, were compared in a specially-bred strain of guinea-pigs. Six litters, I–VI, were available, and were used as *blocks* to take out genetic variation. Increases in weight over a period are given in Table 18.1 in suitable units; the animal from litter IV which should have received diet D showed signs of disease and was removed from the experiment before the final recording.

Table 18.1   Increases in weights of guinea-pigs: data of Example 18.2.1

Litter	I	II	III	IV	V	VI	
Diet A	24	34	41	27	36	32	
B	35	38	46	33	37	35	
C	40	44	54	38	46	40	
D	29	35	40	—	38	34	$T' = 176$
				$B' = 98$			$G' = 856$

In this experiment $r = 6$, $t = 4$. Thus

$$x = \frac{6 \times 98 + 4 \times 176 - 856}{5 \times 3} = \frac{436}{15} = 29 \cdot 07.$$

Since all the other measurements are whole numbers, x will be taken to the nearest whole number, 29; this value is entered in the empty space in the table, and the analysis now carried out as in Example 16.2.1.

For a **Latin Square**, a single lost observation will affect one row, one column and one treatment: the incomplete totals of these are written R', C', T' and the incomplete grand total G'. Then

$$x = \frac{t(R' + C' + T') - 2G'}{(t-1)(t-2)}.$$

A missing plot estimation does not in any sense *recover* the information which has been lost; its purpose is only to allow the simple form of analysis, characteristic of orthogonal designs, to be maintained with the minimum of adjustment. Adjustment of the degrees of freedom has already been mentioned, and in the expression for a significant difference $t_{(f)}$ must have the adjusted value for f, since the d.f. for t are always those for $\hat{\sigma}^2$. The square root $\sqrt{2\hat{\sigma}^2/r}$ in the expression for significant difference will not need to be adjusted as long as two unaffected treatments, such as A and B in Example 18.2.1, are being compared, but when the treatment which suffers from a missing observation (D in the example) is being compared with any of the others we use

$$\sqrt{\hat{\sigma}^2\left(\frac{1}{r}+\frac{1}{r-1}\right)}$$

instead. A warning should be given that the significance levels of the F- and t-tests made when observations are missing do not prove to be exactly 5%, 1% or 0·1%, but something very near to these, and this mathematical disturbance can be neglected so long as the results of significance tests are not interpreted too rigidly.

When more than one plot is lost in randomised blocks or Latin Squares, no simple formulae for the missing values can be given, since there are several possible patterns: the blanks in the table of results might all be in the same block, or all in different blocks, or anything between these extremes, and similarly with respect to treatments. General methods of estimation are given in more advanced texts. The fact that missing values *can* be estimated should never be made an excuse for allowing them to arise, because some precision is lost whenever there are any missing observations, since $t_{(f)}$, $\hat{\sigma}^2$ and the square root are all affected in the expression for significant differences.

18.3 Non-normality

In Chapter 15, we listed the conditions that must be satisfied by the terms in our mathematical models if the analyses described in this book are to be valid (page 141). One condition was that the residual (experimental-error) term e_{ij} should follow a normal distribution, with a variance σ^2 that is constant over all the unit-plots or items used in the experiment. There are two ways in which this requirement may be violated: we may know that the unit-plots, if treated all alike, would produce records which follow some other common distribution, such as the binomial or Poisson, rather than the normal; or it may be clear that some of the treatments induce a more variable response than others. Examples of the latter situation are when the responses to some treatments are much larger than the responses to others, and the responses become more variable as they get larger. We have seen (page 148) how to test whether variances are constant or not, for the different treatments; the problem now is whether we can do anything to make analysis still possible without excessive complication when variances are *not* constant.

The statistical approach to this problem is to analyse, not the original observations, but some mathematical **transformation** of them, such as their square roots or their logarithms; the aim is to find a transformation to a new scale of measurement upon which *the variance will be constant* for all the experimental units. We summarise here the ways of treating some common cases. When data follow the *Poisson* distribution, use the square roots of the observed records, i.e. $\sqrt{y_{ij}}$ instead of y_{ij} in the analysis; for the *binomial*, the *angular transformation* of Fisher and Yates'[2] Table X is appropriate;

this should therefore be applied to data expressed as proportions or percentages. If it seems that variability increases proportionately to the size of response, there are two common cases: should the *variance* seem to increase at the same rate as the size of response, the square root is used, for equality of mean and variance is a property of the Poisson distribution; or if the *standard deviation* increases at the same rate as the size, log y is appropriate. This logarithmic transformation often applies to counts of numbers of insects per leaf or per plant; to remove the difficulty of dealing with log 0 (which is not defined), log $(y + 1)$ may be used—that is, 1 is added to all y-values before taking logarithms.

When transformations have been made, all analyses and significance tests must be carried out on the new (transformed) scale of measurement. This will give statistically sound answers, though unfortunately these new scales are not easy to appreciate: having made significance tests to answer the questions set or test the hypotheses laid down, it is best to rely on verbal explanation of the results, and if tables of means, etc. are needed, a qualified statistician should be asked to help prepare them. There is nothing magic about transformations: their purpose is simply to make the basic models used valid for as many types of data as possible, thus applying statistical methods developed for the normal distribution and not having to redevelop methods for numerous other distributions.

18.4 Transformations in MINITAB

Logarithmic and square-root transformations are possible. The original data are stored in a column, and the logarithm of each can be calculated by the mathematical instruction to place the logs of column 1 into column 2; then the data analysis is done on column 2 rather than column 1. In the same way the square roots could be found and placed in another column. Graphs of the resulting data can be useful; for example the values in a regression problem can be plotted with either y or log y against x, or when studying the *residuals* from a regression or a designed experiment the (hoped-for) improvement in their pattern by changing to a suitable transformation of the data for analysis can be examined graphically.

18.5 Exercises on Chapter 18

18.5.1   In the data of Exercise 16.8.2, suppose that the plant receiving compound A in block V died before the experiment was completed (for a reason not due to treatment), but the remaining 29 observations were exactly as given in that exercise. Estimate a missing value for this entry in the table, and explain how the rest of the analysis would be carried out.

18.5.2   In the data of Exercise 17.5.2, suppose that the entry 13 in block V were lost. Estimate a missing value to replace it, and state what effect this would have on the rest of the analysis.

18.5.3   An organisation which trains technicians has a standard test which is applied at the beginning and at the end of a training course. The difference in the scores on this test is taken as a measure of the success of the course. Four different training schemes A–D are compared on a group of 40 people all at the same stage of training. The data are given below. Use box-and-whisker diagrams to assess whether a standard Analysis of Variance of these data would be a valid way to compare the tests.

A:	33, 44, 39, 38, 29, 41, 39, 30, 42, 44, 26, 33
B:	31, 33, 40, 34, 31, 41, 34, 28, 25
C:	32, 39, 29, 34, 41, 27, 26, 43, 25, 35
D:	42, 46, 42, 42, 46, 39, 43, 41, 38

18.5.4 In a completely randomised experiment, three treatments A, B, C are used, each in eight replicates, and a measurement is made which takes the following values.

A:	12, 8, 15, 10, 17, 19, 11, 16
B:	32, 20, 28, 37, 25, 22, 29, 34
C:	49, 60, 59, 44, 40, 52, 46, 48

Do these data require a transformation before they can be used in an Analysis of Variance? If so, which one?

18.5.5 Apply Bartlett's test (page 148) to the data used in Example 15.3.1 (page 142), and comment on the result.

A guide to choosing a significance test

As explained in Chapter 9, calculating confidence intervals may often give more useful information than carrying out a significance test. However, the basic theoretical methods used are the same, and also most biological literature at present concentrates on significance testing. These notes are therefore written in terms of significance tests, but the obvious parallels in confidence intervals may be made by referring to Chapter 9. Simple tests are based on specific hypotheses about the central tendency (mean or median) of a distribution or about its scatter (variance).

Central tendency of a distibution
(a) Is/are the sample(s) of observations large? If so, the sample mean is a satisfactory basis for a test. See page 71 and Test 8.4.4 for examining the N.H. 'mean $= \mu$'; if the variance σ^2 is not known use s^2 instead. The original distribution of x need not be known, but samples can be smaller if x has a fairly symmetrical distribution. See page 74 and Test 8.6.2 for examining the N.H. 'difference between two means is $\mu_1 - \mu_2$'.
(b) If samples are not large, is the original x distribution normal (or very close to it)? If so, is the variance σ^2 known or must it be estimated by s^2 from the sample? For a single sample from a normal distribution with known variance σ^2, see Test 8.4.1 for examining the N.H. 'mean $= \mu$', and for two samples from normal distributions with known variances see page 73 for a test of the N.H. 'difference between two means is $\mu_1 - \mu_2$'.

   For a single sample from a normal distribution with unknown variance, see Test 8.7.1 for examining the N.H. 'mean $= \mu$'. For two samples from normal distributions with the *same* (unknown) variance, see page 77 and Test 8.7.3 for examining the N.H. 'difference in means is $\mu_1 - \mu_2$' (where $\mu_1 - \mu_2$ is often 0). For several samples from normal distributions with the same (unknown) variance, use Analysis of Variance methods (Chapters 15, 16). If variances are different, see Test 8.7.5.
(c) If samples are not large and the original x distribution is not normal, test hypotheses based on medians rather than means, by nonparametric methods.

   For a single sample, Test 10.2.1 examines the N.H. 'median $= M$', and for two samples Test 10.4.1 examines the N.H. that the two medians are equal; Test 10.2.3 applies when ranks only (not measurements) are available.

Variability in a distribution
(a) Are samples from normal distributions? If not, no simple test is available.
(b) For one sample from a normal distribution, see Test 11.3.1 for examining the

N.H. 'variance $= \sigma^2$'. For two samples from normal distributions, see Test 11.5.1 for examining the N.H. 'variances in both distributions are the same'.

For several samples from normal distributions, see page 148 and Example 15.6.1 for a test (Bartlett's) of whether all variances are the same. Bartlett's test is not very sensitive, nor is it robust to non-normality of the data; other methods may be found in advanced texts, and computer packages often provide *normal probability plots*.

Tests on proportions
See page 89 for testing a hypothesis about a single proportion, and pages 90 and 114 for alternative methods of comparing two proportions.

References and Bibliography

I References

1. RYAN, B.F., JOINER, B.L. and RYAN, T.A. (1985). *MINITAB Handbook*, 2nd edn. PWS-Kent Publishing Company, Boston.
2. FISHER, R. A. and YATES, F. (1963). *Statistical Tables for Biological, Agricultural and Medical Research*, 6th edn. Oliver and Boyd, Edinburgh.
3. CLARKE, G.M. and COOKE, D. (1992). *A Basic Course in Statistics*, 3rd edn. Arnold, London.
4. SIEGEL, S. (1956). *Nonparametric Statistics for the Behavioural Sciences*. McGraw-Hill, New York and Maidenhead.
5. FISHER, R.A. (1950). *Statistical Methods for Research Workers*, 11th edn. Oliver and Boyd, Edinburgh.
6. CAUSTON, D.R. (1983). *A Biologist's Basic Mathematics*. Arnold, London.
7. CLARKE, G.M. and KEMPSON, R.E. (1997). *Introduction to the Design and Analysis of Experiments*. Arnold, London.
8. DUNCAN, D.B. (1955). *Biometrics*, 11, 1–42.
9. EDWARDS, A.L. (1968). *Experimental Design in Psychological Research*. Holt, Rinehart and Winston, New York.
10. COCHRAN, W.G. and COX, G.M. (1992). *Experimental Designs*, 2nd edn. Wiley, New York and Chichester.

II Bibliography

For a detailed study of experimental design and analysis, the book by Clarke and Kempson, or that by Cochran and Cox, mentioned above should be consulted.

A classical text on the methods and applications of statistics is: SNEDECOR, G.W. and COCHRAN, W.G. (1963). *Statistical Methods*, 6th edn. Iowa State College Press.

Those whose interests are in the area of medical statistics should consult: ARMITAGE, P. and BERRY, G., *Statistical Methods in Medical Research*. Blackwell, Oxford. This is also a very useful general text.

Answers

Chapter 2
2.7.1 Frequencies of $r = 0, 1, 2, 3, 4, 5$ are 3, 13, 30, 33, 17, 4. Display in a bar-diagram.
2.7.2 Histogram using interval end-points as 49·5, 79·5, . . . , 179·5, 239·5. Areas represent frequencies; note the unequal widths of intervals.
2.7.3 Frequencies A: (0) 13, (1) 8, (2) 5, (3) 5, (4) 5, (5) 2, (6) 5, (7) 0, (8) 3, (9) 1, (10) 2, (22) 1. B: (0) 15, (1) 8, (2) 5, (3) 5, (4) 8, (5) 3, (6) 3, (7) 1, (15) 1, (22) 1. Bar diagrams *or* dot-plots on the *same* x-scale for frequency, directly underneath one another for comparison.
2.7.4 Frequency table

Weight (g)	15–19	20–24	25–29	30–34	35–39	40–44	45–49	Total
Frequency	9	6	15	11	8	8	3	60

An ordered stem-and-leaf diagram using these intervals gives maximum information. Alternatives are histogram, dot-plot.
2.7.5 Frequencies in intervals are (a) 4, 5, 23, 58, 61, 30, 3, 3; (b) 4, 4, 6, 30, 46, 54, 29, 9, 3, 2. (b) is rather better as it gives more intervals with the frequencies spread out more.

Chapter 3
3.10.1 An ordered stem-and-leaf diagram is useful for finding median and quartiles. $n = 36$, mean = 20·86, median = 21, quartiles are 19 and 23, variance = 10·0087, standard deviation = 3·164.
3.10.2 $n = 38$, mean = 21·32, median = 21, quartiles are 19 and 24, variance = 13·3030, standard deviation = 3·647. Because the two new values are above all the others, the mean increases; so does the variance because the values are some distance away from most. Median and quartiles little affected.
3.10.3 Mean = 30·08, standard deviation = 8·448, median = 30, quartiles are 24·5 and 36·5. Median near middle of box, and whiskers of similar length, indicating general symmetry.
3.10.4 Median is 94th in order, which is 46. Quartiles are 47th from bottom and 47th from top, which are 42 and 49. The cumulative frequency curve 2.7.5(a) has upper end-points to intervals as 19·5, 33·5, . . . , 61·5, 65.5 and cumulative frequencies must be plotted against these as x-values.
3.10.5 Mean and standard deviation are *not* suitable because the data are very skew. For A, median = 2, quartiles are 0 and 5; for B, median = 2, quartiles are 0 and 4. Medians equal, semi-inter-quartile ranges $\frac{5}{2} = 2\cdot5$ and $\frac{4}{2} = 2$; B rather less scattered.
3.10.6 Mean = 52·33, s.d. = 4·677. First six: mean = 48·83, s.d. = 3·312. Second six: mean = 55·83, s.d. = 2·787. Median = 52·5, quartiles are 48·5, 56·5, semi-inter-quartile range = 4. First six: median = 48·5, quartiles are 47 and 52, semi-inter-quartile range = 2·5. Second six: median = 56·5, quartiles are 53 and 58, semi-inter-quartile range = 2·5.
3.10.7 Mid-points (x) of intervals are 64·5, 84·5, 94·5, 104·5, 114·5, 124·5, 139·5, 164·5, 209·5.

Mean $= 118\cdot29$, variance $= 645\cdot3764$, s.d. $= 25\cdot404$, median $= 109\cdot5 + \frac{13\cdot5}{24} \times 10 = 115\cdot125$. Mean slightly higher because of the small number of very large observations in the last interval.

3.10.8 Mean $= 1\cdot12$, variance $= 9\cdot0873$. Unsuitable because data are highly skew.

Chapter 4

4.7.2 (a) Number the members 001–750, read digits in runs of three, neglect 000 and 751–999 and any repeats. (b) Number the members 001–250, read digits in runs of three, subtract 250 from all between 251 and 500, subtract 500 from all between 501 and 750, subtract 750 from all between 751 and 000 (taking 000 as 1000). Then use remainders to locate members, ignoring any repeats. (c) As in (b), except that 301–600 and 601–900 now correspond to 001–300; 901–999, 000 and any repeats are ignored. (d) Take runs of four digits; first two give distance in one direction (01–25) and others in other direction (01–40). Set 26–50, 51–75 and 76–00 to correspond to 01–25; and 41–80 to correspond to 01–40 (rest of final pairs cannot be used).

4.7.3 Stratify into faculties. Decide how many to take from each: may be 20 each, or slightly more from Arts than from Science so that proportion from each is about the same. Number the Arts students 001–363 and the Science students 001–285, and select two separate independent samples as in Exercise 4.7.2. Need two strata because level of expenditure may be different.

4.7.4 (a) Larger households are more likely to be selected (provided they contain more registered voters). Better to compile a list of properties (houses, flats etc.) and sample at random from this; possibly stratify into different areas, streets or ages of building. (b) Children of larger families are less likely to be selected; either make up a list of children or, when sampling, return the card to the pack if there are more children (still to be selected) in a family. (c) Systematic sample only satisfactory if there are no trends of phase 10 plants running through the plantation (e.g. unwise if rows contain (say) 40 plants as we may obtain too many, or too few, end plants).

Chapter 5

5.9.1 (a) $\frac{9}{16} = 0\cdot5625$. (b) $\frac{3}{4} = 0\cdot75$. (c) $\frac{3}{16} \times \frac{3}{16} = 0\cdot0352$. (d) $2 \times \frac{9}{16} \times \frac{1}{16} = 0\cdot0703$.

5.9.2 $\frac{1}{10} \times \frac{1}{5} = 0\cdot02$, or 2%. Probability of 100 'failures' $= 0\cdot98^{100} = 0\cdot1326$. Mean $= np = 100 \times 0\cdot02 = 2$.

5.9.3 $n = 8$, $p = 0\cdot6$ in binomial. $P(6) + P(7) + P(8) = 0\cdot3154$.

5.9.4 Mean $\bar{r} = \frac{1}{120} ((6 \times 0) + (31 \times 1) + (42 \times 2) + (29 \times 3) + (10 \times 4) + (2 \times 5)) = \frac{252}{120} = 2\cdot10$. Using $\bar{r} = n\hat{p}$ gives $2\cdot10 = 5\hat{p}$, so $\hat{p} = 0\cdot42$.

5.9.5 (a) All conditions for binomial satisfied; $n = 12$, $p = \frac{1}{3}$. (b) This is a geometric distribution. (c) Not binomial because this is a 'finite' population; probability changes after each selection. (d) Not binomial because p should be changing as 'learning' takes place, run-by-run and session-by-session. (e) Binomial, $n = 4$, $p = \frac{1}{2}$.

5.9.6 $n = 5$, $p = 0\cdot8$ for satisfactory items. $P(\text{accept}) = P(5) = (0\cdot8)^5 = 0\cdot328$; $P(\text{reject}) = P(0) + P(1) + P(2) = 0\cdot058$; $P(\text{another sample}) = P(3) + P(4) = 0\cdot614$. In 10 items, nine satisfactory if there were four satisfactory first time and then five (second sample only taken when first showed three or four satisfactory). $P(4) \times P(5) = 0\cdot4096 \times 0\cdot328 = 0\cdot1342$. We assume all items have the same probability of failure, independently of all others.

5.9.7 (a) $\frac{6!}{4!2!} \left(\frac{1}{2}\right)^6 = 0\cdot2344$. (b) $P(4) + P(5) + P(6) = (15 + 6 + 1) \left(\frac{1}{2}\right)^6 = 0\cdot343\ 75$; excess may be of *either* sex, so total probability $= 0\cdot6875$.

5.9.8 $P(2\ \text{normal}) = \left(\frac{3}{4}\right)^2 = \frac{9}{16} = 0\cdot5625$.

5.9.9 The counts follow a geometric distribution. Mean count $= \frac{194}{50} = 3\cdot88$; estimate of p is $\frac{1}{3\cdot88} = 0\cdot258$ (i.e. 25·8%), provided all sampled fruits have the same probability of being Grade I independently of all the others.

Chapter 6

6.5.1 (a) $P(0) = e^{-1} = 0.3679$. (b) $P(1) = 1 \times e^{-1} = 0.3679$. (c) *Either* there are 0, *or* there is 1, *or* there are 2 or more. $P(2 \text{ or more}) = 1 - P(0) - P(1) = 0.2642$.

6.5.2 $\bar{r} = \frac{70}{100} = 0.7$. Variance $= \frac{1}{99}\left\{\sum fr^2 - (\sum fr)^2 / 100\right\} = \frac{1}{99}(394 - 49) = \frac{345}{99} = 3.485$. Mean and variance are very different, which indicates the Poisson is unsuitable (mean = variance in Poisson). Values 5, 8, 15 indicate some 'grouping'.

6.5.3 $\bar{r} = 2.1$. Variance $= 2.475$. The difference between mean and variance is small, so a Poisson model is acceptable, which suggests that the plants may well be 'growing at random'.

6.5.4 When $\mu = 1$, $P(\text{not more than } 1) = P(0) + P(1) = 0.7358$. Second part requires $P(5)$ in binomial with $n = 10$, $p = 0.7358$; this is $\frac{10!}{5!5!}(0.7358)^5(0.2642)^5 = 0.070$.

6.5.5 From the data, the estimate of $P(0)$ is $\frac{10}{50} = 0.2$. Solving $e^{-\mu} = 0.2$ gives $\mu = 1.609$.

6.5.6 Mean $= 2.41$. Expect $8.98(0)$, $21.65(1)$, $26.08(2)$, $20.95(3)$, $12.62(4)$, 9.72 (5 or more). Agreement not bad. Variance $= 2.1029$, reasonably near to mean.

6.5.7 Mean $= 2$. Expect $6.77(0)$, $13.53(1)$, $13.53(2)$, $9.02(3)$, $4.51(4)$, $2.64(5 \text{ or more})$. Variance $= 4.0408$. Agreement bad; too many large r observed.

6.5.8 (a) $\mu = 3.3$, so (i) $P(4) = e^{-3.3}(3.3)^4 / 4! = 0.1823$; (ii) $P(0) + P(1) + P(2) + P(3) = e^{-3.3}(1 + 3.3 + (3.3)^2/2! + (3.3)^3/3!) = 0.036\ 88 \times 15.7345 = 0.5803$; (iii) *either* we have 4, *or* less than 4, *or* more than 4, so this probability is $1 - 0.1823 - 0.5803 = 0.2374$. Poisson conditions of randomness, independence, constant rate often apply to insect counts. Mean per five minutes $= 1.1$; $P(0) = e^{-1.1} = 0.3329$. (b) (i) $P(0) + P(1) + P(2)$ in binomial with $n = 10$ and $p = 0.3329$ is $0.003\ 37$; (ii) $np = 3.33$.

Chapter 7

7.10.1 (a) $z = \frac{(5.00 - 3.95)}{\sqrt{2.25}} = \frac{1.05}{1.5} = 0.70$; (b) $z = \frac{(0.29 - 0.50)}{\sqrt{0.64}} = -\frac{0.21}{0.8} = -0.26$; (c) $z = \frac{(-0.47 - 1.38)}{\sqrt{1.21}} = -\frac{1.85}{1.1} = -1.68$; (d) $z = \frac{[-6.89 - (-6.50)]}{\sqrt{0.04}} = -\frac{0.39}{0.2} = -1.95$.

7.10.2 (a) Same mean; second more scattered, s.d. twice that of first. (b) Same scatter; second a long way to the right of first because mean much greater. (c) Same scatter; second a long way to left of first, mean is eight units smaller. (d) Second has mean slightly larger, variation (hence scatter) much smaller. When drawing, use fact that almost all of normal curve lies between mean plus and minus three standard deviations.

7.10.3 $\bar{x}$ is $N(10, \frac{25}{64})$, i.e. $N(10, 0.390\ 625)$. For $\bar{x} = 9$, $z = \frac{(9 - 10)}{\sqrt{0.390\ 625}} = \frac{-1}{0.625} = -1.60$. $P(z < -1.60) = P(z > +1.60) = 1 - P(z < +1.60) = 1 - 0.9452 = 0.0548$ (from Table VII).

7.10.4 Distribution $N(2.43, 0.025^2)$. For $x = 2.40$ and 2.41, values of z are $z = \frac{(2.40 - 2.43)}{0.025} = -1.20$ and $z = \frac{(2.41 - 2.43)}{0.025} = -0.80$. The probability that z is between -1.20 and -0.80 is (by symmetry) the same as the probability that z is between $+0.80$ and $+1.20$. From Table VII, $P(0.80 < z < 1.20) = P(z < 1.20) - P(z < 0.80) = 0.8849 - 0.7881 = 0.0968$. Average of 10 is distributed $N\left(2.43, \frac{0.025^2}{10}\right)$, i.e. $N(2.43, 0.000\ 062\ 5)$; the s.d. is $0.007\ 91$. For $x = 2.44$, $x = \frac{(2.44 - 2.43)}{0.007\ 91} = 1.265$; and for $x = 2.43$, $z = 0$. $P(z < 1.265) = 0.8971$, and $P(z < 0) = 0.5000$. The required probability is 0.3971.

7.10.5 S.d. of $\bar{x}$ is $\sigma/\sqrt{n}$; we require $\frac{0.5}{\sqrt{n}} = 0.1$, so $n = 25$.

7.10.6 Mean is multiplied by 0.4536 to convert to kg, and variance by $(0.4536)^2$; distribution becomes $N(4.536, 3.0143)$.

7.10.7 Mean mass $= 455.51$ g; distribution is $N(455.51, 33.33)$. Probability that mass is less than 454 is $P\left(z < \frac{(454 - 455.51)}{\sqrt{33.33}}\right) = P\left(z < -\frac{1.51}{5.773}\right) = P(z < -0.2616) = P(z > +0.2616)$ by symmetry $= 1 - 0.6032 = 0.3968$ from Table VII.

7.10.8 Variances of both normal distributions (fuel, labour time). If σ_F^2, σ_L^2 are variances for fuel and labour, total cost is normal with mean $25\,C_F + 4C_L$ where C_F, C_L are the unit costs for fuel and labour; variance is $C_F^2\sigma_F^2 + C_L^2\sigma_L^2$.

7.10.9   Bag weights are $N(5\cdot1, (0\cdot05)^2)$; mean of 10 bags is $N\left(5\cdot1, \frac{(0\cdot05)^2}{10}\right) = N(5\cdot1, 0\cdot000\ 25)$.

7.10.10   Standard method $N(13, (1\cdot2)^2)$. $P(x > 15\cdot5) = P\left(z > \frac{(15\cdot5 - 13\cdot0)}{1\cdot2}\right) = P(z > 2\cdot083) = 1 - P(z < 2\cdot083) = 0\cdot0186$ from Table VII. New method $N(14\cdot2, (1\cdot2)^2)$. $P(x > 15\cdot5) = P\left(z > \frac{(15\cdot5 - 14\cdot2)}{1\cdot2}\right) = P(z > 1\cdot083) = 1 - P(z < 1\cdot083) = 0\cdot1393$. Increase in proportion of Grade I blooms is $0\cdot1393 - 0\cdot0186 = 0\cdot1207$ (i.e. 12·07% of the production).

7.10.11   (a) No; n much too small. (b) $N(80 \times \frac{3}{5}, 80 \times \frac{3}{5} \times \frac{2}{5}) = N(48, 19\cdot2)$. (c) No; larger n needed with p as small as this. (d) Although p is near 1, n is very large: $N\left(300 \times \frac{9}{10}, 300 \times \frac{9}{10} \times \frac{1}{10}\right) = N(270, 27)$.

7.10.12   Form of original distribution not given but unless it is very skew $\bar{x}$ will be *approximately* $N(5, \frac{8}{200}) = N(5, 0\cdot04)$. If only 20 observations available, nothing can be said unless distribution of x is known. (If it is itself normal, the mean is exactly $N(5, \frac{8}{20})$.)

7.10.13   Estimated $\hat{p} = \frac{120}{500} = 0\cdot24$. Normal approximation to r is $N(100 \times 0\cdot24, 100 \times 0\cdot24 \times 0\cdot76) = N(24, 18\cdot24)$. $P(r < 14\cdot5)$, using the continuity correction, is $P\left(z < \frac{(14\cdot5 - 24)}{\sqrt{18\cdot24}}\right) = P\left(z < -\frac{9\cdot5}{4\cdot27}\right) = P(z < -2\cdot22) = 0\cdot0131$.

7.10.14   Binomial, $n = 396$, $p = \frac{72}{396} = 0\cdot1818$. Normal approximation may be used: $N(72, 72 \times 0\cdot8182) = N(72, 58\cdot9091)$. One-quarter of 396 is 99, and one-third is 132. The z values corresponding to these are $\frac{(99-72)}{\sqrt{58\cdot9091}} = \frac{27}{7\cdot675} = 3\cdot52$, and $\frac{(132-72)}{\sqrt{58\cdot9091}} = 7\cdot82$; both of these are so large that the probability between them is negligible.

7.10.15   The lower 4% corresponds in $N(0, 1)$ to $z = -1\cdot75$, and the upper $22\frac{1}{2}\%$ to $z = +0\cdot755$ (from Table VII). Because $z = \frac{(x-\mu)}{\sigma}$, $x = \mu + \sigma z$ is the value of x corresponding to a given z. When $z = -1\cdot75$, $x = 99$; and when $z = +0\cdot755$, $x = 104$. Therefore $\mu - 1\cdot75\sigma = 99$ and $\mu + 0\cdot755\ \sigma = 104$. Solving these gives $\mu = 102\cdot49$ and $\sigma = 1\cdot996$.

Chapter 8

8.10.1   (a) $z = \frac{(4\cdot34 - 2\cdot75)}{\sqrt{1\cdot64}} = \frac{1\cdot59}{1\cdot28} = 1\cdot24$; do not reject N.H. (b) $z = \frac{(11\cdot28 - 15\cdot30)}{\sqrt{2\cdot72}} = -\frac{4\cdot02}{1\cdot65} = -2\cdot44$; reject N.H. in favour of A.H. as stated. (c) $z = \frac{(-1\cdot03 - (-0\cdot55))}{\sqrt{0\cdot096}} = -\frac{0\cdot48}{0\cdot31} = -1\cdot55$; do not reject N.H. (d) $z = \frac{(21\cdot92 - 18\cdot70)}{\sqrt{3\cdot24}} = \frac{3\cdot22}{1\cdot80} = 1\cdot79$; reject N.H. because a one-tail test is needed for the given A.H. (and z is greater than 1·645). (e) $z = \frac{(1\cdot45 - 3\cdot00)}{\sqrt{0\cdot35}} = -\frac{1\cdot55}{0\cdot59} = -2\cdot62$; reject N.H. in favour of given A.H. (in a one-tail test). (f) The A.H. cannot be accepted in preference to the N.H. for any values of μ that are greater than 0, as this one is, because the A.H. cannot explain such values better than the N.H.

8.10.2 $\bar{x}$ is $N(10, \frac{25}{64})$ on the N.H., i.e. $N(10, 0\cdot390\ 625)$. For $\bar{x} = 11\cdot1$, $z = \frac{(11\cdot1 - 10\cdot0)}{0\cdot625} = 1\cdot76$. Do not reject the N.H., as no specific A.H. is given and the test is therefore two-tail.

8.10.3   On the N.H., $\bar{x}$ is $N(29\cdot8, \frac{25}{20}) = N(29\cdot8, 1\cdot25)$. For $\bar{x} = 28$, $z = \frac{(28 - 29\cdot8)}{\sqrt{1\cdot25}} = -\frac{1\cdot8}{1\cdot118} = -1\cdot61$; do not reject the N.H.

8.10.4 $\bar{x}$ is approximately $N(5, \frac{8}{200})$ on the N.H., this is $N(5, 0\cdot04)$. For $\bar{x} = 4\cdot77$, $z = \frac{(4\cdot77 - 5\cdot00)}{\sqrt{0\cdot04}} = -\frac{0\cdot23}{0\cdot2} = -1\cdot15$; do not reject the N.H.

8.10.5   Binomial, $n = 500$, $p = 0\cdot9$ on N.H. Approximate by $N(450, 45)$. For $r = 425$, $z = \frac{(425 - 450)}{\sqrt{45}} = -\frac{25}{6\cdot708} = -3\cdot73$. Strong evidence to reject N.H. in favour of A.H. that p is not 0·9.

8.10.6   Since there are 50 sampling units, the normal approximation to the mean of a Poisson distribution can reasonably be applied although μ is less than 5: $\bar{r}$ is approximately $N(2, \frac{2}{50})$ on the N.H. For $\bar{r} = 2\cdot2$, $z = \frac{(2\cdot2 - 2\cdot0)}{\sqrt{0\cdot04}} = \frac{0\cdot2}{0\cdot2} = 1\cdot0$; do not reject the N.H.

8.10.7   On the N.H., x is $N(110, 84)$, so $\bar{x}$ is $N\left(110, \frac{84}{10}\right)$. We are not told that the distribution of yield is normal, but this is reasonable for crop data and we cannot proceed without it. For

$\bar{x} = 114$, $z = \frac{114-110}{\sqrt{8\cdot4}} = \frac{4\cdot0}{2\cdot898} = 1\cdot38$; there is not enough evidence to reject the N.H. in favour of an A.H. of improved yield.

8.10.8 Binomial, $n = 200$, $p = 0\cdot8$ on N.H.; approximate the distribution of r by $N(160, 32)$. For $r = 170$, $z = \frac{170-160}{\sqrt{32}} = 1\cdot77$, and in a one-tail test of the A.H. that p is greater than $0\cdot8$ there is evidence to reject the N.H. in favour of the A.H. Neither (a) nor (b) can lead to this conclusion because both appear less successful.

8.10.9 Assuming weights are normally distributed, the N.H. is that $\bar{x}$ is $N\left(102, \frac{49}{100}\right)$; for $\bar{x} = 99$, $z = \frac{(99-102)}{\sqrt{0\cdot49}} = -\frac{3}{0\cdot7} = -4\cdot29$, very strong evidence to reject the N.H. in favour of an A.H. that $\bar{x}$ is not 102 (and the result is so strong that it is not critical whether the x-distribution is exactly normal). With five children, normality *would* be important; replacing 100 by 5 leads to a non-significant result in the same test.

8.10.10 Small sample, so assumption of normal distribution needed. Sample mean $\bar{x} = 93\cdot11$, variance $s^2 = 85\cdot3521$, $n = 10$, $\mu = 100$ on N.H. $t_{(9)} = \frac{(93\cdot11-100\cdot00)}{\sqrt{85\cdot3521/10}} = -\frac{6\cdot89}{2\cdot922} = -2\cdot36$, significant at 5% and providing evidence to reject N.H. in favour of A.H. that mean is not 100. (If normality not assumed, see Chapter 10.)

8.10.11 Treat the differences between pairs as a single sample of eight observations and take as N.H. that the mean difference is 0. The differences should be a normally distributed set; there is one large outlying value which casts doubt on this. $\bar{d} = 4\cdot0$ and $s_d^2 = 18\cdot5714$. $t_7 = \frac{(4\cdot0-0)}{\sqrt{18\cdot5714/8}} = \frac{4}{1\cdot524} = 2\cdot63$, significant at 5%, giving evidence against the N.H.

8.10.12

r:	1	2	3	4	5	6	7	8	9
Frequency:	0	2	5	19	60	57	32	3	2
Cumulative F:	0	2	7	26	86	143	175	178	180

Median value (between 90th and 91st observation in rank order) is 6. Mean $= 5\cdot572$. Variance $= 1\cdot3523$; s.d. $= 1\cdot163$. Little to choose between mean and median as data fairly symmetrical.

Large sample: $z = \frac{5\cdot572-5}{\sqrt{1\cdot3523/180}} = \frac{0\cdot572}{0\cdot0867} = 6\cdot60$; reject N.H. that $\mu = 5$, in favour of the A.H. that μ is not 5.

8.10.13 $\hat{p}_1 = \frac{32}{60} = 0\cdot5333$; $p_2 = \frac{41}{105} = 0\cdot3905$. $\frac{\hat{p}_1(1-\hat{p}_1)}{n_1} + \frac{\hat{p}_2(1-\hat{p}_2)}{n_2} = \frac{(0\cdot5333\times0\cdot4667)}{60} + \frac{(0\cdot3905\times0\cdot6095)}{105} = 0\cdot006\,414\,858$, and its square root is $0\cdot0801$. $z = \frac{(0\cdot5333-0\cdot3905)}{0\cdot0801} = 1\cdot78$, which does not provide sufficient evidence to reject a N.H. of equal attack (in favour of an A.H. that the proportions are unequal).

8.10.14 Assumption of normally distributed weights needed. A: mean $= 97\cdot0$, $n = 11$, variance $= 847\cdot2$; C: mean $= 56\cdot0$, $n = 11$, variance $= 774\cdot8$. Ratio of variances $= 1\cdot09$, F-test does not suggest inequality so can pool variances to give $s^2 = 811\cdot0$ with 20 d.f. N.H. is that mean comb size is the same for A and C. $t_{(20)} = \frac{(97\cdot0-56\cdot0)}{\sqrt{811\cdot0(\frac{1}{11}+\frac{1}{11})}} = \frac{41\cdot0}{12\cdot14} = 3\cdot38$; reject N.H.

8.10.15 Sample mean $= 0\cdot4$, sample variance $= 0\cdot6229$. $t = \frac{(0\cdot4-0\cdot1)}{\sqrt{0\cdot6229/8}} = \frac{0\cdot3}{0\cdot28} = 1\cdot07$; do not reject the given N.H.

8.10.16 F-test of variances: $\frac{15\cdot2824}{8\cdot0275} = 1\cdot90$, not significant as $F_{(10, 15)}$ and the variances can be pooled. $s^2 = \frac{[(10\times15\cdot2824)+(15\times8\cdot0275)]}{25} = 10\cdot9295$. $t_{(25)} = \frac{(6\cdot65-4\cdot28)}{\sqrt{10\cdot9295(\frac{1}{11}+\frac{1}{16})}} = \frac{2\cdot37}{1\cdot295} = 1\cdot83$; do not reject N.H. that means are the same.

8.10.17 $n = 12$, $\mu = 6\cdot0$, $\bar{x} = 5\cdot75$, $s^2 = 0\cdot44$. Assuming a normal distribution of differences, $t_{(11)} = \frac{(5\cdot75-6\cdot0)}{\sqrt{0\cdot44/12}} = \frac{-0\cdot25}{0\cdot101} = 1\cdot31$; do not reject N.H. that true mean distance is $6\cdot0$.

Chapter 9
9.6.1 Significance test of N.H. '$\mu = 4$' is $t_{(24)} = \frac{(5\cdot85-4\cdot0)}{\sqrt{4\cdot84/25}} = \frac{1\cdot85}{0\cdot44} = 4\cdot20$; reject this N.H. Confidence interval is from $5\cdot85 - 2\cdot064 \times 0\cdot44$ to $5\cdot85 + 2\cdot064 \times 0\cdot44$ at 95%, i.e. $(4\cdot94$ to $6\cdot76)$.

9.6.2   Mean $= 18 \cdot 67$, $n = 6$, sample variance $= 19 \cdot 867$. $t_{(5)} = 2 \cdot 571$. 95% confidence limits are $18 \cdot 67 \pm 2 \cdot 571 \sqrt{\frac{19 \cdot 867}{6}}$ i.e. $18 \cdot 67 \pm 2 \cdot 571 \times 1 \cdot 820$, i.e. $18 \cdot 67 \pm 4 \cdot 68$ or $(13 \cdot 99$ to $23 \cdot 35)$.

9.6.3 $\hat{p} = 0 \cdot 85$, $n = 200$. Interval $0 \cdot 85 \pm 1 \cdot 96 \sqrt{0 \cdot 85 \times 0 \cdot 15/200} = 0 \cdot 85 \pm 0 \cdot 05 = (0 \cdot 80$ to $0 \cdot 90)$.

9.6.4   Confidence limits for mean weight of 100 children are found assuming the variance is 49. Then the lower limit is $99 - 1 \cdot 96 \sqrt{\frac{49}{100}} = 99 - 1 \cdot 96 \times 0 \cdot 7 = 99 - 1 \cdot 37 = 97 \cdot 63$, and the upper limit is $99 + 1 \cdot 37 = 100 \cdot 37$ at the 95% level. At 99%, $99 \pm 2 \cdot 576 \times 0 \cdot 7$ gives $(97 \cdot 20, 100 \cdot 80)$.

9.6.5   (a) $\mu \pm 1 \cdot 96\sigma$, i.e. $100 \pm 1 \cdot 96\sqrt{10}$ or $(93 \cdot 8$ to $106 \cdot 2)$. (b) S.d. is $\sqrt{10/10}$ instead of $\sqrt{10}$: $100 \pm 1 \cdot 96$ or $(98 \cdot 0$ to $102 \cdot 0)$. (c) S.d. is now $\sqrt{10/100}$: $100 \pm 1 \cdot 96 \sqrt{0 \cdot 1}$ or $(99 \cdot 4$ to $100 \cdot 6)$. Without knowing σ, (a) is not possible; (b) calculate sample variance and proceed using $t_{(9)}$ instead of z; (c) as (b) except that z can be used because the sample is very large.

9.6.6   Normal approximation to a Poisson mean $N(4, \frac{4}{50}) = N(4, 0 \cdot 08)$. Approximate 95% limits are $4 \pm 1 \cdot 96 \sqrt{0 \cdot 08}$, i.e. $(3 \cdot 45$ to $4 \cdot 55)$.

9.6.7   95% confidence limits to d are $4 \cdot 0 \pm t_{(7)} \times 1 \cdot 524$ (see Exercise 8.10.11), i.e. $4 \cdot 0 \pm 2 \cdot 365 \times 1 \cdot 524$, or $4 \cdot 0 \pm 3 \cdot 60$, or $(0 \cdot 4$ to $7 \cdot 6)$.

9.6.8   First sample: $n = 9$, $\bar{x} = 25 \cdot 78$, $s^2 = 13 \cdot 4119$. Interval $25 \cdot 78 \pm t_{(8)}\sqrt{13 \cdot 4119/9} = 25 \cdot 78 \pm 2 \cdot 82 = (22 \cdot 96$ to $28 \cdot 60)$. Second sample: $n = 11$, $\bar{x} = 28 \cdot 24$, $s^2 = 5 \cdot 2525$. Interval $28 \cdot 24 \pm t_{(10)}\sqrt{5 \cdot 2525/11} = 28 \cdot 24 \pm 1 \cdot 54 = (26 \cdot 70$ to $29 \cdot 78)$. Difference (Second $-$ First) $= 2 \cdot 46 \pm t_{(18)} \sqrt{s^2(\frac{1}{9} + \frac{1}{11})}$ with pooled $s^2 = 8 \cdot 8789$, gives interval $(-0 \cdot 35$ to $+ 5 \cdot 27)$. Variances not significantly different.

9.6.9 $\hat{p} = \frac{172}{250} = 0 \cdot 688$. Assuming same p, $1 \cdot 96 \sqrt{\frac{(0 \cdot 688 \times 0 \cdot 312)}{n}}$ is to equal $0 \cdot 03$, i.e. $0 \cdot 214\,656/n = (0 \cdot 03/1 \cdot 96)^2 = 0 \cdot 000\,234\,28$ which gives $n = 916 \cdot 25$; say $n = 920$.

9.6.10 $n_1 = 250$, $\hat{p}_1 = \frac{172}{250} = 0 \cdot 688$, $\frac{\hat{p}_1(1 - \hat{p}_1)}{n_1} = 0 \cdot 000\,858\,6$, $n_2 = 200$, $\hat{p}_2 = \frac{158}{200} = 0 \cdot 790$, $\frac{\hat{p}_2(1 - \hat{p}_2)}{n_2} = 0 \cdot 000\,829\,5$. Variance of difference $(p_1 - p_2)$ is $0 \cdot 000\,858\,6 + 0 \cdot 000\,829\,5 = 0 \cdot 001\,688\,1$, and its square root is $0 \cdot 0411$. Observed difference $= 0 \cdot 790 - 0 \cdot 688 = 0 \cdot 102$, and 95% limits are $0 \cdot 102 \pm 1 \cdot 96 \times 0 \cdot 411$, i.e. $0 \cdot 102 \pm 0 \cdot 081$ or $(0 \cdot 021$ to $0 \cdot 183)$.

9.6.11   Differences are $0 \cdot 17$, $-0 \cdot 07$, $0 \cdot 59$, $0 \cdot 06$, $0 \cdot 07$, $0 \cdot 11$, $-0 \cdot 32$, $0 \cdot 04$; pairing removes systematic species differences but the resulting figures do not look to be a sample from a normal distribution, so the results may be affected. Mean difference $= 0 \cdot 081\,25$, variance $= 0 \cdot 064\,812\,5$. $n = 8$. $t_{(7)} = 2 \cdot 365$ (5%), $3 \cdot 499$ (1%). $\sqrt{s^2/n} = 0 \cdot 090$. Limits are $0 \cdot 081\,25 \pm 2 \cdot 365 \times 0 \cdot 090$ i.e. $0 \cdot 081\,25 \pm 0 \cdot 213$ or $(-0 \cdot 132$ to $0 \cdot 294)$ at 95%; and $0 \cdot 081\,25 \pm 0 \cdot 315$ or $(-0 \cdot 234$ to $0 \cdot 398)$ at 99%.

Chapter 10

10.6.1   No readings equal to $7 \cdot 8$, so can use all $n = 25$. Number above $7 \cdot 8$ is binomial with $p = \frac{1}{2}$, approximate by $N(25 \times \frac{1}{2}, 25 \times \frac{1}{2} \times \frac{1}{2}) = N(12 \cdot 5, 6 \cdot 25)$. Observed number greater than $7 \cdot 8$ is $17 \cdot 5$, so $z = \frac{(18 - 12 \cdot 5)}{\sqrt{6 \cdot 25}} = 5 \cdot 0/2 \cdot 5 = 2 \cdot 0$; reject hypothesis that readings are equally likely to be above or below $7 \cdot 8$, i.e. median $= 7 \cdot 8$, in favour of A.H. that median is not $7 \cdot 8$. Sample mean $= 7 \cdot 804$, so any test will accept $7 \cdot 8$ as a value for the mean; however, the data have some outliers, and appear not to be normally distributed, so a t-test cannot be used.

10.6.2   Joint ranking is ABAABABABAABAABAABABBBABABABAAABAAA. $\frac{1}{2} n_1 n_2 = 120$; $\frac{1}{12} n_1 n_2 (n_1 + n_2 + 1) = 660$. Number of times B comes before A is $U = 132$. $z = \frac{(132 - 120)}{\sqrt{660}} = 12 \cdot 0/25 \cdot 7 = 0 \cdot 467$, no evidence to reject a N.H. of equal medians.

10.6.3   Only $n = 22$ (out of 30) people stated a preference. A sign test for $n = 22$, $n_+ = 15$ gives $z = \frac{(14 \cdot 5 - 11)}{\sqrt{5 \cdot 5}} = \frac{3 \cdot 5}{2 \cdot 345} = 1 \cdot 49$, giving no reason (in a two-tail test) to reject the N.H. that there is no difference.

10.6.4 Person differences w_i are $-1, 0, 1, 4, 5, 0, -1, 0, 2, 1, 0, 4, 5, -2, 0, 6, -3, 1, 0, -1, 2,$ $-3, 0, -1, 1, 3, 3, 0, 1, 4$. Ranks for the absolute values $|w_i|$, ignoring zeroes, are $5, 5, 18, 20\frac{1}{2},$ $5, 11, 5, 18, 20\frac{1}{2}, 11, 22, 14\frac{1}{2}, 5, 5, 11, 14\frac{1}{2}, 5, 5, 14\frac{1}{2}, 14\frac{1}{2}, 5, 18$. Since there are nine values of w equal to ± 1, they share the first nine places in the rank order, and all carry average rank 5. Other tied ranks are dealt with in the same way. $R_+ = 193$, $R_- = 60$, so $T = 60$. On the N.H. of no difference, T is $N(126 \cdot 5, 948 \cdot 75)$, so $z = \frac{(60 - 126 \cdot 5)}{\sqrt{948 \cdot 75}} = \frac{-66 \cdot 5}{30 \cdot 80} = -2 \cdot 16$, providing evidence to reject the N.H.

10.6.5 (a) A one-tail test is now required; $z = 1 \cdot 49$ is not significant at the 5% level. (b) One-tail significance values are required but the results are the same as before.

10.6.6 Differences:

(B − A)	0·7	0·2	1·0	−0·4	−0·5	1·6	0·8	−0·3	−0·3	0·8	0·1	−0·1	0·7	0·3	1·3	−0·1	−0·3	−0·3
Ranks	12½	4	16	10	11	18	14½	7	7	14½	2	2	12½	7	17	2	7	7

(a) $R_+ = 118$, $R_- = 53$, so $T = 53$. On the N.H. of no difference, T is normal with mean $= \frac{1}{4} n(n + 1) = \frac{(18 \times 19)}{4} = 85 \cdot 5$ and variance $\frac{1}{24} n(n + 1)(2n + 1) = \frac{(18 \times 19 \times 37)}{24} = 527 \cdot 25$. So $z = \frac{(53 - 85 \cdot 5)}{\sqrt{527 \cdot 25}} = -\frac{32 \cdot 5}{22 \cdot 96} = -1 \cdot 42$, which gives no evidence for rejecting the N.H. (b) Mean $\bar{d} = 0 \cdot 289$, $s_d^2 = 0 \cdot 4081$, N.H. is 'true mean $d = 0$'. $t_{(17)} = \frac{(0 \cdot 289 - 0)}{\sqrt{0 \cdot 4081/18}} = \frac{0 \cdot 289}{0 \cdot 151} = 1 \cdot 92$, not significant, so result is the same as before. Dot-plot etc. may be used to examine whether the differences are normal; no obvious outliers.

Chapter 11

11.6.1 First variety: $n = 10$, $s^2 = 689 \cdot 433$. Second variety: $n = 12$, $s^2 = 225 \cdot 061$. F-test of N.H. that $\sigma_1^2 = \sigma_2^2$: $\frac{689 \cdot 433}{255 \cdot 061} = 2 \cdot 70$ is $F_{(9, 11)}$ and is not significant at 5% so N.H. is not rejected. Confidence interval for σ_1^2 uses $\chi_{(9)}^2$, lower and upper $2\frac{1}{2}\%$ points being $2 \cdot 700$ and $19 \cdot 023$. $(n-1) s^2/\chi^2 = 9 \times 689 \cdot 433/\chi^2 = 6204 \cdot 90/\chi^2$, giving limits $(326 \cdot 18$ to $2298 \cdot 11)$. For σ_2^2, lower and upper points of $\chi_{(11)}^2$ are $3 \cdot 816, 21 \cdot 92$. $(n-1) s^2 = 11 \times 255 \cdot 061 = 2805 \cdot 67$, so limits are $\frac{2805 \cdot 67}{21 \cdot 92}, \frac{2805 \cdot 67}{3 \cdot 816}$ i.e. $(128 \cdot 00$ to $735 \cdot 24)$.

11.6.2 $s_I^2 = 13 \cdot 969$; $s_{II}^2 = 18 \cdot 488$. Testing N.H. '$\sigma^2 = 15$': for I, $\frac{(9 \times 13 \cdot 969)}{15} = 8 \cdot 38$, not significant as $\chi_{(9)}^2$; for II, $\frac{(9 \times 18 \cdot 488)}{15} = 11 \cdot 09$, also not significant as $\chi_{(9)}^2$, so the N.H. cannot be rejected in either case. Upper and lower $2\frac{1}{2}\%$ points of $\chi_{(9)}^2$ are $2 \cdot 700, 19 \cdot 023$. Confidence intervals: I, $9 \times 13 \cdot 969/\chi^2$, i.e. $\frac{125 \cdot 721}{\chi^2}$, giving $(6 \cdot 61$ to $46 \cdot 56)$; II, $\frac{9 \times 18 \cdot 488}{\chi^2} = \frac{166 \cdot 392}{\chi^2}$, giving $(8 \cdot 75$ to $61 \cdot 63)$. Dot-plots not very informative for small samples; but there are no very extreme outlying values.

11.6.3 $F_{(8, 10)} = \frac{13 \cdot 4119}{5 \cdot 2525} = 2 \cdot 55$, which is not significant at 5%, so it was acceptable to use the pooled variance in Exercise 9.6.8. Combining both calculations, there is no evidence of difference either in mean weight or in variability of weight due to the two soil types.

Chapter 12

12.7.1 $\chi^2 = 2 \cdot 88$ with 4 d.f. Model fits well.

12.7.2

	Cured	Not cured	
A	172 (183·33)	78 (66·67)	250
B	158 (146·67)	42 (53·33)	200
	330	120	450

For N.H. of no difference in proportion cured, $\chi_{(1)}^2 = \frac{(-11 \cdot 33)^2}{183 \cdot 33} + \frac{(11 \cdot 33)^2}{66 \cdot 67} + \frac{(11 \cdot 33)^2}{146 \cdot 67} + \frac{(-11 \cdot 33)^2}{53 \cdot 33} = 5 \cdot 91$, significant at 5% providing evidence to reject the N.H.

12.7.3 All expected frequencies will be 80. $\chi^2_{(9)} = \frac{(85-80)^2}{80} + \frac{(77-80)^2}{80} + \cdots + \frac{(84-80)^2}{80} + \frac{(77-80)^2}{80} = \frac{306}{80} = 3\cdot825$, providing no evidence against the N.H. of equal frequencies.

12.7.4

	a	b	c	d	Total
I	75 (66)	15 (24)	25 (21)	5 (9)	120
II	85 (88)	37 (32)	26 (28)	12 (12)	160
III	60 (66)	28 (24)	19 (21)	13 (9)	120
	220	80	70	30	400

Expected values: $66 = \frac{(220 \times 120)}{400}$, $24 = \frac{(80 \times 120)}{400}$, etc. $\chi^2_{(6)} = \sum \frac{(O-E)^2}{E} = \frac{(75-66)^2}{66} + \cdots + \frac{(13-9)^2}{9} = 11\cdot35$, not significant, giving no evidence against N.H. of genetic equivalence.
For strain II,

	a	b+c+d	Total
OBS	85	75	160
EXP	90	70	160

$\chi^2_{(1)} = \frac{(85-90)^2}{90} + \frac{(75-70)^2}{70} = 0\cdot635$, giving no evidence against a N.H. of $a:(b+c+d) = 9{:}7$.

12.7.5

		Aster		
		√	×	Total
Atriplex	√	29 (25·83)	12 (15·17)	41
	×	34 (37·17)	25 (21·83)	59
		63	37	100

Expected values: $\frac{(63 \times 41)}{100} = 25\cdot83$, etc. $\chi^2_{(1)} = (3\cdot17)^2 \left\{ \frac{1}{25\cdot83} + \frac{1}{15\cdot17} + \frac{1}{37\cdot17} + \frac{1}{21\cdot83} \right\} = 1\cdot78$, not significant so no evidence against N.H. of independence.

12.7.6

	1937–46	1947	Total
Age 0–14	467 (423·2)	453 (496·8)	920
15 and over	131 (174·8)	249 (205·2)	380
	598	702	1300

N.H. is of no age difference in attack. Expected frequencies: $\frac{(598 \times 920)}{1300} = 423\cdot2$, etc.
$\chi^2_{(1)} = (43\cdot8)^2 \left\{ \frac{1}{423\cdot2} + \frac{1}{174\cdot8} + \frac{1}{496\cdot8} + \frac{1}{205\cdot2} \right\} = 28\cdot20$, very strong evidence against the N.H.
Accept that there *was* an age difference.

12.7.7

	FN	FP	CL	CP	Total
O	328	122	77	33	560
E	315	105	105	35	560

$\chi^2_{(3)} = \frac{(328-315)^2}{315} + \frac{(122-105)^2}{105} + \frac{(77-105)^2}{105} + \frac{(33-35)^2}{35} = 10\cdot87$, significant at 5%, providing evidence against the 9:3:3:1 hypothesis.

12.7.8

	P	G	C	Total
O	55 (47·6)	21 (23·8)	162 (166·6)	238
A	37 (43·0)	24 (21·5)	154 (150·5)	215
B	8 (9·4)	5 (4·7)	34 (32·9)	47
	100	50	350	500

$\chi^2_{(4)} = \frac{(55-47·6)^2}{47·6} + \cdots + \frac{(34-32·9)^2}{32·9} = 3·08$, providing no evidence against N.H. of the same distribution of types over each blood group.

12.7.9 Should be $\frac{1}{4}$ ebony if the N.H. is true.

	Ebony	Not ebony	Total
OBS:	27	63	90
EXP:	22·5	67·5	90

$\chi^2_{(1)} = (4·5)^2 \left(\frac{1}{22·5} + \frac{1}{67·5}\right) = 1·20$, no evidence against the N.H.

12.7.10 Total observed frequency = 120. Probabilities are $P(r) = e^{-1·8} (1·8)^r/r!$, and these are $(r=0)$ 0·1653, (1) 0·2975, (2) 0·2678, (3) 0·1607, (4) 0·0723; hence $P(r \geqslant 5) = 0·0364$. Multiply by 120 for expected frequencies.

r	0	1	2	3	4	$\geqslant 5$	Total
OBS:	27	35	24	18	8	8	120
EXP:	19·84	35·70	32·14	19·28	8·68	4·37	120

χ^2 has 4 d.f. (lose 1 for total, 1 for estimated mean). $\chi^2_{(4)} = \frac{(27-19·84)^2}{19·84} + \cdots + \frac{(8-4·37)^2}{4·37} = 7·82$, which is not significant at 5% and so provides no evidence against the Poisson hypothesis.

12.7.11 Final interval must be '5 or more' in order to avoid small frequencies. $\bar{r} = \frac{1}{120}$ - $((0 \times 26) + (1 \times 30) + (2 \times 26) + (3 \times 18) + (4 \times 9) + (5 \times 5) + (6 \times 6)) = 233/120 = 1·94$. Calculate probabilities with $\mu = 1·94$ in the Poisson formula: $(r=0)$ 0·1435, (1) 0·2783, (2) 0·2700, (3) 0·1746, (4) 0·0847, so that $P(r \geqslant 5)$ is 0·0489. Multiply by 120 for expected frequencies.

r	0	1	2	3	4	$\geqslant 5$	Total
OBS	26	30	26	18	9	11	120
EXP	17·22	33·40	32·40	20·95	10·16	5·87	120

$\chi^2_{(4)} = \frac{(26-17·22)^2}{17·22} + \cdots + \frac{(11-5·87)^2}{5·87} = 11·12$, significant at 5% and providing evidence against the N.H. of a Poisson model. (d.f. are four because the mean was not given, so had to be estimated). There seem to be too many 0s and too many observations in the top cell, suggesting some grouping or lack of independence.

12.7.12 Binomial, $n=3$ and $p=\frac{1}{6}$, has probabilities $(r=0)$ 0·5787, (1) 0·3472, (2) 0·0694, (3) 0·0046. Multiply by 216 to obtain expected frequencies: 2 and 3 must be combined because otherwise 3 would have a very small expected frequency.

r	0	1	(2, 3)	Total
OBS:	110	85	21	216
EXP:	125	75	16	216

$\chi^2_{(2)} = \frac{(110-125)^2}{125} + \frac{(85-75)^2}{75} + \frac{(21-16)^2}{16} = 4·70$ which is not significant at 5%, so the N.H. (of the given binomial) is not rejected.

Chapter 13

13.7.1 $\sum x_i = 2142$, $\sum y_i = 1482$, $n = 15$, $\sum x_i^2 = 312\,562$, $\sum y_i^2 = 154\,818$, $\sum x_i y_i = 211\,007$.

$$r = \frac{(15 \times 211\,007 - 2142 \times 1482)}{\sqrt{(15 \times 312\,562 - 2142^2)(15 \times 154\,818 - 1482^2)}} = \frac{-9339}{\sqrt{100\,266 \times 125\,946}} = \frac{-9339}{112\,375} = -0.08, \text{ not significant.}$$

13.7.2 $\sum x_i = 1050$, $\sum y_i = 991$, $n = 15$, $\sum x_i^2 = 73\,560$, $\sum y_i^2 = 65\,509$, $\sum x_i y_i = 69\,384$.

$$r = \frac{(15 \times 69\,384 - 1050 \times 991)}{\sqrt{(15 \times 73\,560 - 1050^2)(15 \times 65\,509 - 991^2)}} = \frac{+210}{\sqrt{900 \times 554}} = \frac{+210}{706} = +0.2974, \text{ not significant.}$$

13.7.3 Leaf area difficult and tedious to measure. Use height instead when comparing different plants for measurements influenced by leaf area. Height is a quick and easy indicator of leaf area in plants of the same variety.

13.7.4 $\sum x_i = 160.4$, $\sum y_i = 79.4$, $\sum x_i^2 = 2981.52$, $\sum x_i y_i = 1429.72$, $\sum y_i^2 = 702.3$. Sum of products $= 1429.72 - \frac{(160.4 \times 79.4)}{9} = 14.6356$, s.s. for $y = 702.3 - \frac{79.4^2}{9} = 1.8156$, s.s. for $x = 2981.52 - \frac{160.4^2}{9} = 122.8356$. $r = \frac{14.6356}{\sqrt{122.8356 \times 1.8156}} = +0.980$.

13.7.5 (a) The value is significant at 0.1% (36 d.f.) This indicates that rainfall decreases as distance inland increases. The statement is in stronger terms than are justified only on a significance test, but probably has truth in it, though it does not imply that *all* rainfall is in coastal strip. (b) The population is heterogeneous (different breeds of sow), and it is known that reproduction depends to some extent on breed; so a single correlation coefficient for the whole population is not informative.

13.7.6

Ranks	A	B	C	D	E	F	G	H	J	K	L	M
I	8	6	3	7	1	9	10	11	4	2	12	5
II	6	4	10	12	1	9	11	3	8	2	5	7
d_i	2	2	−7	−5	0	0	−1	8	−4	0	7	−2

$n = 12$. $\sum d_i^2 = 216$. $r_s = 1 - \{(6 \times 216)/(12 \times 143)\} = 0.2448$.
Rank correlation more suitable when data not jointly normally distributed: neither set of percentages here looks obviously normal. Also if any actual values are numerically wrong, ranking may reduce the effect of this, as in E (Method I) which is ranked 1 whether 89 or 59 is the correct figure. Ranking particularly useful where one or two observations are very different size from others, so dominating an ordinary correlation (see 13.5).

13.7.7

Ranks	I	II	III	IV	V	VI	VII	VIII
X	2	1	5	7	3	6	4	8
Y	4	5	1	7	3	6	2	8
d_t	−2	−4	4	0	0	0	2	0

$n = 8$. $\sum d_i^2 = 40$. $r_s = 1 - \{(6 \times 40)/(8 \times 63)\} = +0.524$. $r_s \sqrt{\frac{n-2}{1-r_s^2}} = 0.524 \sqrt{\frac{6}{1-0.2744}} = 0.524 \times 2.876 = 1.51$, which does not approach significance when tested as $t_{(6)}$. The five small pairs of records (I, II, III, V, VII) are in reverse rank order, while the other three pairs are in the same order. In calculating r, the largest observations, numerically, dominate the calculation to give a highly significant value, 0.996; but the sizes X, Y are clearly not normally distributed so the significance of r cannot be tested. There are really two distinct populations of environments.

13.7.8 Nothing, since we have neither continuous data nor ranks, so cannot use r or r_s.

Chapter 14

14.7.1 $\sum x_i = 16.5$, $\sum y_i = 209$, $n = 6$, $\therefore \bar{x} = 2.75$, $\bar{y} = 34.83$. $\sum x_i^2 = 66.25$, $\sum y_i^2 = 9941$, $\sum x_i y_i = 810$, $\therefore \sum (x_i - \bar{x})^2 = 20.875$, $\sum (x_i - \bar{x})(y_i - \bar{y}) = 235.25$, $\sum (y_i - \bar{y})^2 = 2660.83$.

$\therefore \hat{b} = 235 \cdot 25/20 \cdot 875 = 11 \cdot 27$, line is $(y - 34 \cdot 83) = 11 \cdot 27 \ (x - 2 \cdot 75)$, i.e. $y = 11 \cdot 27x + 3 \cdot 84$. So for $x = 0, \frac{1}{2}, 1, 2, 3, 4, 5, 6$, we predict $y = 3 \cdot 84, 9 \cdot 48, 15 \cdot 11, 26 \cdot 38, 37 \cdot 65, 48 \cdot 92, 60 \cdot 19, 71 \cdot 46$. The value $y = 3 \cdot 84$ at $x = 0$ may represent base-level of nutrients present in soil; *or* the relation between y and x may become a curved one when x approaches 0. S.S. for regression $= (235 \cdot 25)^2/20 \cdot 875 = 2651 \cdot 14$, S.S. for deviations $= 2660 \cdot 83 - 2651 \cdot 14 = 9 \cdot 69$, with 4 d.f., $\hat{\sigma}^2 = 9 \cdot 69/4 = 2 \cdot 4225$, $F_{(1, \ 4)} = 2651 \cdot 14/2 \cdot 4225 = 1094$, significant at $0 \cdot 1\%$ $\therefore$ line good fit. Var $(\hat{b}) = 2 \cdot 4225/20 \cdot 875 = 0 \cdot 116$, s.e.$(\hat{b}) = 0 \cdot 34$. $\therefore$ 95% limits to b are $\hat{b} \pm t_{(4, \ 0 \cdot 05)}$ s.e. $(\hat{b})$ $= 11 \cdot 27 \pm 2 \cdot 776 \times 0 \cdot 34$ i.e. $10 \cdot 33$ to $12 \cdot 21$.

14.7.2 $n = 7$. $y = $ content, $x = $ temperature. $\sum x = 16 \cdot 8$, $\sum y = 38 \cdot 5$, $\sum x^2 = 62 \cdot 32$, $\sum y^2 = 245 \cdot 49$, $\sum xy = 119 \cdot 36$, $\hat{b} = \frac{119 \cdot 36 - (16 \cdot 8)(38 \cdot 5)/7}{62 \cdot 32 - (16 \cdot 8)^2/7} = \frac{26 \cdot 96}{22 \cdot 0} = 1 \cdot 225$. $y - \bar{y} = \hat{b}(x - \bar{x})$. $\bar{x} = 2 \cdot 40$, $\bar{y} = 5 \cdot 50$, so line is $y - 5 \cdot 50 = 1 \cdot 225 \ (x - 2 \cdot 40) = 1 \cdot 225x - 2 \cdot 94$. i.e. $y = 1 \cdot 225x + 2 \cdot 56$. Points on this line are $(x = 2, \ y = 5 \cdot 01)$ and $(x = 5, \ y = 8 \cdot 685)$. SS$_{\text{Regression}} = \frac{26 \cdot 96^2}{22 \cdot 00} = 33 \cdot 038 \ 25$.

Total SS$(y) = 245 \cdot 49 - \frac{38 \cdot 5^2}{7} = 33 \cdot 74$ $\therefore$ $s^2 = \frac{1}{5}(33 \cdot 74 - 33 \cdot 038 \ 25) = 0 \cdot 140 \ 35$. Var $(\hat{b})$ $= \frac{s^2}{22 \cdot 00} = 0 \cdot 006 \ 380$. $F_{(1, \ 5)} = 235 \cdot 4$ very highly significant, therefore reject N.H. '$b = 0$'. For use in confidence interval, $t_{(5)}$ at 5% (two tail) $= 2 \cdot 571$. 95% limits $1 \cdot 225 \pm 2 \cdot 571 \times \sqrt{0 \cdot 006 \ 38}$, i.e $1 \cdot 225 \pm 0 \cdot 205$: $(1 \cdot 02$ to $1 \cdot 43)$.

14.7.3 Graph suggests fitting log $y = a + bx$. For $x = 0, 1, 2, \ldots, 8$, values of log y are $-0 \cdot 12$, $0 \cdot 08$, $0 \cdot 24$, $0 \cdot 40$, $0 \cdot 54$, $0 \cdot 67$, $0 \cdot 79$, $0 \cdot 92$, $1 \cdot 06$, with mean $0 \cdot 51$. $\bar{x} = 4 \cdot 0$. $\sum(x_i - \bar{x})^2 = 60$. $\sum(x_i - \bar{x})(Y_i - \bar{Y}) = 8 \cdot 61$. ($Y$ stands for log y). Thus $\hat{b} = 8 \cdot 61/60 = 0 \cdot 144$. $\therefore$ log $y - 0 \cdot 51 = 0 \cdot 144 \times (x - 4)$, i.e. log $y = 0 \cdot 144x - 0 \cdot 067$.

14.7.4 Line is $y = 3 \cdot 849x - 0 \cdot 536$ (some suggestion of slight curve at two ends but line fits well). If 10 added to all y-values, a increases by 10 and b does not change. If all y increased by 9%, both a and b increase by 9%.

Chapter 15

15.9.1 Method totals: A, $9 \cdot 10$; B, $9 \cdot 25$; C, $9 \cdot 05$. Grand total $= 27 \cdot 40$. Total S.S. $= 0 \cdot 009 \ 533$. Methods S.S. $= 0 \cdot 004 \ 333$. $\therefore$ Residual S.S. $= 0 \cdot 005 \ 200$. Methods M.S. (2 d.f.) $= 0 \cdot 002 \ 17$, Residual M.S. (12 d.f.) $= 0 \cdot 000 \ 43$. $F_{(2, \ 12)} = 5 \cdot 00$, significant at 5%. Means: A, $1 \cdot 82$; B, $1 \cdot 85$; C, $1 \cdot 81$. B v. A: $\frac{1 \cdot 85 - 1 \cdot 82}{\sqrt{2 \times 0 \cdot 000 \ 43/5}} = \frac{0 \cdot 03}{\sqrt{0 \cdot 000 \ 173}} = \frac{0 \cdot 03}{0 \cdot 013} = 2 \cdot 28$, significant at 5% as $t_{(12)}$.

C v. A: not significant.

15.9.2 Total S.S. $= 2508 \cdot 72$. Varieties S.S. $= 1208 \cdot 47$. Residual S.S. $= 1300 \cdot 25$. Varieties M.S. (7 d.f.) $= 172 \cdot 64$, Residual M.S. (24 d.f.) $= 54 \cdot 18$, $F_{(7, \ 24)} = 3 \cdot 19$, significant at 5%. Means: A, $37 \cdot 3$; B, $35 \cdot 0$; C, $39 \cdot 0$; D, $30 \cdot 8$; E, $26 \cdot 8$; F, $34 \cdot 0$; G, $48 \cdot 0$; H, $30 \cdot 0$. Significant differences are $t_{(24)} \sqrt{2 \times 54 \cdot 18/4} = t_{(24)} \sqrt{27 \cdot 09} = t_{(24)} \times 5 \cdot 2$. Values of $t_{(24)}$ are $2 \cdot 064$ (5%), $2 \cdot 797$ (1%), $3 \cdot 745$ ($0 \cdot 1\%$), so significant differences are $10 \cdot 7$ (5%), $14 \cdot 5$ (1%), $19 \cdot 5$ ($0 \cdot 1\%$). Suggests G better than B (significant at 5%), E worse than C (significant at 5%), but variability high and more replicates desirable.

15.9.3 P: $r = 6$, total 1867; Q: 6, 2117; R: 6, 1839; S: 6, 2287; C: 12, 3017. Treatment S.S. $= (1867^2 + 2117^2 + 1839^2 + 2287^2)/6 + 3017^2/12 - 11 \ 127^2/36 = 82 \ 632$. Total S.S. $= 3 \ 596 \ 855 - 11 \ 127^2/36 = 157 \ 685$. Residual S.S. $= 75 \ 053$. D.f. are 35 for Total, 4 for Treatments, 31 for Residual. Residual M.S. $= 2421 = s^2$. S.E. of difference between mean of C and mean of any other $= \sqrt{s^2(\frac{1}{6} + \frac{1}{12})} = 24 \cdot 60$. Means: P, $311 \cdot 2$; Q, $352 \cdot 8$; R, $306 \cdot 5$; S, $381 \cdot 2$; C, $251 \cdot 4$. $t_{(31)} = 2 \cdot 04$ (5%), $2 \cdot 75$ (1%), $3 \cdot 64$ ($0 \cdot 1\%$). P, R differ from C at 5% level; Q, S differ from C at $0 \cdot 1\%$ level. S.E. of difference $(R - S) = \sqrt{s^2(\frac{1}{6} + \frac{1}{6})} = 28 \cdot 41$. Observed difference $= -74 \cdot 7$. 95% limits are $-74 \cdot 7 \pm 2 \cdot 04 \times 28 \cdot 41 = -74 \cdot 7 \pm 58 \cdot 0$; interval is $(-132 \cdot 7$ to $-16 \cdot 7)$.

15.9.4 Totals A, 119; B, 128; C, 93; O, 201; G $= 541$, sum of squares of all 25 observations is $25^2 + 23^2 + \ldots + 22^3 + 17^2 = 12 \ 009$, and total S.S. $= 12 \ 009 - 541^2/25 = 301 \cdot 76$. Treatments S.S. $= \frac{1}{5}(119^2 + 128^2 + 93^2) + \frac{1}{10}(201^2) - \frac{1}{25}(541^2) = 171 \cdot 66$. D.F. are 3 for treatments, 24 for total,

21 for residual. Residual S.S. $= 301{\cdot}76 - 171{\cdot}66 = 130{\cdot}10$, M.S. $= 6{\cdot}20 = s^2$. Treatments M.S. $= 57{\cdot}22$. $F_{(3,\,21)} = 9{\cdot}23^{***}$, least sig. diff. to compare O with A, B or $C = \sqrt{s^2\left(\frac{1}{10} + \frac{1}{5}\right)} \times t_{(21)} = t\sqrt{1{\cdot}86} = 1{\cdot}364\,t$. Values of $t_{(21)}$ are $2{\cdot}080$ (5%), $2{\cdot}831$ (1%), $3{\cdot}819$ (0·1%), so sig. diffs. are $2{\cdot}85$ (5%), $3{\cdot}88$ (1%), $5{\cdot}23$ (0·1%). Means are: A, 23·8; B, 25·6; C, 18·6; O, 20·1, so A, B appear to give higher growth than O (5% and 0·1% levels respectively).

Chapter 16

16.8.1   Total S.S. $= 1{\cdot}3783$, Blocks S.S. $= 0{\cdot}5233$, Treatments (Densities) S.S. $= 0{\cdot}7083$, $\therefore$ Residual S.S. $= 0{\cdot}1467$. Blocks M.S. (5 d.f.) $= 0{\cdot}1047$, Treatments M.S. (3 d.f.) $= 0{\cdot}2361$, Residual M.S. (15 d.f.) $= 0{\cdot}0098$. For Blocks, $F_{(5,\,15)} = 10{\cdot}70$, significant at 0·1%, for treatments $F_{(3,\,15)} = 24{\cdot}15$, significant at 0·1%. Blocks have removed a considerable amount of systematic variation. For treatment means, the significant differences are $t_{(15)}\,\sqrt{2\hat{\sigma}^2/6} = t\,\sqrt{0{\cdot}0033}$. $t_{(15)} = 2{\cdot}131$ at 5%, 2·947 at 1%, 4·073 at 0·1%, and $\sqrt{0{\cdot}0033} = 0{\cdot}057$. So significant differences are 0·12 (5%), 0·17 (1%), 0·23 (0·1%). Means are A: 2·82; B: 3·00; C: 3·28; D: 3·13. Steady increase, which is significant, from A to C, then drop to D.

16.8.2   Total S.S. $= 2055{\cdot}47$, Blocks S.S. $= 163{\cdot}13$, Compounds S.S. $= 1720{\cdot}67$, $\therefore$ Residual S.S. $= 171{\cdot}67$. Blocks M.S. (4 d.f.) $= 40{\cdot}78$, Compounds M.S. (5 d.f.) $= 344{\cdot}13$, Residual M.S. (20 d.f.) $= 8{\cdot}58$. For Blocks, $F_{(4,\,20)} = 4{\cdot}75$, significant at 1%, for Compounds $F_{(5,\,20)} = 40{\cdot}11$, significant at 0·1%. $\sqrt{2\hat{\sigma}^2/r} = \sqrt{2 \times 8{\cdot}58/5} = \sqrt{3{\cdot}432} = 1{\cdot}85$. $t_{(20)} = 2{\cdot}086$ (5%), 2·845 (1%), 3·850 (0·1%), so significant differences are 3·9 (5%), 5·3 (1%), 7·1 (0·1%). B, D, F not significantly different, others higher than these.

16.8.3   Taster totals a 83, b 84, c 85, d 84. Sessions 76, 84, 85 and 91. Suppliers A 94, B 78, C 78, D 86. So $G = 336$. S.S. for tasters $= (83^2 + 84^2 + 85^2 + 84^2)/4 - 336^2/16 = 0{\cdot}5$, with 3 d.f. S.S. for sessions $= (76^2 + 84^2 + 85^2 + 91^2)/4 - 336^2/16 = 28{\cdot}5$, with 3 d.f. S.S. for suppliers $= (94^2 + 78^2 + 78^2 + 86^2)/4 - 336^2/16 = 44{\cdot}0$ with 3 d.f. Total S.S. $= 21^2 + \ldots + 26^2 - 336^2/16 = 78{\cdot}0$, with 15 d.f. Hence $S_E = 5{\cdot}0$, with 6 d.f., and $\hat{\sigma}^2 = 5{\cdot}0/6 = 0{\cdot}833$. M.S. for tasters, sessions, suppliers are 0·167, 9·500, 14·667, and $F_{(3,\,6)} < 1$ (n.s.) for tasters, $F_{(3,\,6)} = 11{\cdot}40^{**}$ for sessions (perhaps due to steady increase in score from session 1 to 4). $F_{(3,\,6)} = 17{\cdot}60^{**}$ for products, indicating an overall difference. Test any specific hypotheses about products in t-tests with 6 d.f.

16.8.4   Block totals 19·6; 20·9; 20·5; 21·0; 21·5; giving S.S. $= 0{\cdot}5050$ (4 d.f.). Density totals A, 21·9; B 24·1; C, 27·5; D, 30·0, giving S.S. $= 7{\cdot}7215$ (3 d.f.). Total S.S. (19 d.f.) $= 10{\cdot}8375$, hence residual M.S. (12 d.f.) $= 0{\cdot}217583$ and $F_{(4,12)}$ for blocks is less than 1, so apparently no need for blocking in future experiments on this site. For densities, $F_{(3,12)} = 11{\cdot}83$, significant at 0·1%. Clearly differences exist, and a regression of growth on density would extract useful information.

16.8.5 $\sigma = 2{\cdot}8$. Residual d.f. will be $(r-1)(t-1) = 21$. $t_{(21,\,0{\cdot}05)} = 2{\cdot}08$. $\delta = 2{\cdot}08 \times 2{\cdot}8\sqrt{2/4} = 4{\cdot}12$. This represents a 22·9% change in mean.

16.8.6   When $V = 8\%$ and $\delta = 10\%$, $10 = 8t\sqrt{2/r}$, and using $t \doteq 2$ this gives $100r = 512$, so $r = 6$ is the minimum needed. Increasing r to 12, with $V = 12\%$, $\delta = 12t\sqrt{2/12} \doteq 24/\sqrt{6} = 9{\cdot}8\%$, about the same as before.

Chapter 17

17.5.1   Totals: Blocks 80, 85, 89, 96; Treatments ① 58, n 104, p 65, np 123. Blocks S.S. $= 34{\cdot}25$ (3 d.f.), N $= 676{\cdot}00$ (1 d.f.), P $= 42{\cdot}25$ (1 d.f.), NP $= 9{\cdot}00$ (1 d.f.), Total S.S. $= 829{\cdot}75$, Residual S.S. $= 68{\cdot}25$ (9 d.f.) $\therefore$ Residual M.S. $= 7{\cdot}58$. Blocks and NP not significant, N significant at 0·1%, P at 5%. Mean $+ N = 28{\cdot}4$, $-N = 15{\cdot}4$, difference significant at 0·1%; $+P = 23{\cdot}5$, $-P = 20{\cdot}3$, difference significant at 5%. (No need for t-tests, since F has only one d.f. and so measures the same thing.)

17.5.2 Blocks S.S. = 26·8 (4 d.f.), Blocks M.S. = 6·7, S.S. for A = 304·2, B = 192·2, AB = 88·2 (each 1 d.f.), Total S.S. = 751·8, Residual S.S. = 140·4 (12 d.f.), Residual M.S. = 11·7. Blocks n.s., A, B, AB all significant. Mean for A alone = 25·0, higher than all others, so AB interaction implies that B counterbalances beneficial effect of A when added together.

17.5.3 Analysis of Variance

Source of Variation	D.F.	S.S.	M.S.	
R	3	112·1250	37·375	
T	2	44·3333	22·167	
RT	6	89·0000	14·833	$F_{(6, 12)} = 3.46*$
Treatments	11	245·4583		
Residual	12	51·5000	$4·2917 = \hat{\sigma}^2$	
Total	23	296·9583		

Interaction significant. Construct two-way table of means. Within this table sig. diffs are $t_{(12)} \sqrt{2 \times 4·2917/2} = 2·072t = 4·51$ (5%), 6·33 (1%), 8·95 (0·1%).

Means	T_1	T_2	T_3
R_1	12·0	15·0	17·5
R_2	18·0	21·5	15·5
R_3	18·0	20·5	19·5
R_4	17·5	19·5	25·0

Can compare T means at each R level, or R means at each T level, to look for reasons for interaction.

17.5.4 Sum of all 48 observations = 1062, sum of squares = 26 722. Total S.S. = 26 722 − 1062²/48 = 3225·25. Add the pairs to give treatment totals:

Temperature	lower			higher		
Variety	B	C	L	B	C	L
Time 1	22	15	30	34	19	35
2	37	27	44	42	31	46
3	46	39	50	59	43	63
4	61	54	65	71	58	71

S.S. for treatments = $\frac{1}{2}(22^2 + 15^2 + \ldots + 58^2 + 71^2) - 1062^2/48 = 2988·25$. Main effect Varieties from totals: B, 372; C, 286; L, 404; S.S. = $\frac{1}{16}$ ($372^2 + 286^2 + 404^2$) − 1062²/48 = 465·50. Temperature totals: Low, 490; High 572. S.S. = $\frac{1}{24}$ ($490^2 + 572^2$) − 1062²/48 = 140·08. Time totals: (1), 155; (2), 227, (3), 300; (4), 380. S.S. = $\frac{1}{12}$ ($155^2 + \ldots + 380^2$) − 1062²/48 = 2332·75.

Totals in two-way tables:

	Low temp.	High temp		Variety	B	C	L			B	C	L
Time 1	67	88		Time 1	56	34	65		Low	166	135	189
2	108	119		2	79	58	90		High	206	151	215
3	135	165		3	105	82	113					
4	180	200		4	132	112	136					

S.S. for Time + Temp + Interaction = $\frac{1}{16}(67^2 + \ldots + 200^2) - 1062^2/48 = 2487·92$, so interaction = 15·09; similarly Time × V. Interaction = 6·00, Temp × V. Interaction = 18·17.

Analysis of Variance.

Source of variation	D.F.	S.S.	M.S.	
Temperatures	1	140·08	140·08	
Times	3	2332·75	777·58	
Varieties	2	465·50	232·75	
Temp × Time	3	15·08	5·03	
Temp × Varieties	2	18·17	9·09	All F-tests against
Time × Varieties	6	6·00	1·00	s^2 non-significant.
Temp × Time × V	6	10·67	1·78	
	23	2988·25		
Residual	24	237·00	$9·875 = s^2$	
Total	47	3225·25		

No evidence of interactions. Summarise by giving means for each main effect. For each temperature separately, draw graph with time as x and percentage as y, using different symbols for each variety: the means of pairs (treatments) are plotted. Shapes of response similar because there are no interactions. All main effects are very highly significant.

Chapter 18

18.5.1 $B'_V = 40$, $T'_A = 105$, $G' = 358$. ∴$x = (5 \times 40 + 6 \times 105 - 358)/(5 \times 4) = 472/20 = 23·6$. Take missing value as 24, put in table of results, reduce residual d.f. to 19, total d.f. to 28.

18.5.2 Treat as randomised block. Incomplete $B'_V = 55$, $T' = 52$, $G' = 305$. ∴$x = 14·8$. Take x as 15, reduce residual d.f. to 11, total to 18. Factorial part of analysis as usual.

18.5.3 For easiest comparison using box-and-whisker diagrams, replication should be the same for every treatment. However, these are not very different: $r_A = 12$, $r_B = 9$, $r_c = 10$, $r_D = 9$, so four diagrams underneath one another on the same scale of measurement may be drawn, as in Chapter 3. The medians and quartiles are: A, $M = 38\frac{1}{2}$, q $= 31\frac{1}{2}$; Q $= 41\frac{1}{2}$; B, $M = 33$, q $= 29\frac{1}{2}$, Q $= 37$; C, $M = 33$, q $= 27$, Q $= 39$; D, $M = 42$, q $= 40$, Q $= 44\frac{1}{2}$. Ranges are: A, 18; B, 16; C, 18; D, 8. B, C and D all seem reasonably symmetrical, but A is rather skew to the left. The range of D is noticeably less than the others. Therefore the assumptions for Analysis of Variance may not be very well satisfied, though no obvious transformation would improve this.

18.5.4 For A, $\bar{y} = 13·5$, $s = 3·82$, $s^2 = 14·5714$; for B, $\bar{y} = 28·4$, $s = 5·88$, $s^2 = 34·5536$; for C, $\bar{y} = 49·8$, $s = 6·99$, $s^2 = 48·7857$. So s^2 is roughly proportional to $\bar{y}$ for these sets of data, which suggests that a square-root transformation would be useful.

18.5.5 For A, $\bar{y} = 20·43$, $s = 4·86$, $s^2 = 23·6190$; for B, $\bar{y} = 41·57$, $s = 6·70$, $s^2 = 44·9524$; for C, $\bar{y} = 31·43$, $s = 5·41$, $s^2 = 29·2857$. In Bartlett's test, $k = 3$, each s_i^2 has $f_i = 6$ d.f., $s^{*2} = 32·6190$. $C = 1 + \frac{1}{6}(\frac{3}{6} - \frac{1}{18}) = 29/27$. $M = 2·3026(18 \times 1·5135 - 6 \times \{1·3733 + 1·6528 + 1·4667\}) = 0·6590$. $M/C = 0·614$, n.s. as $\chi^2_{(2)}$. But s^2 seems roughly proportional to $\bar{y}$ and so a square-root transformation might improve the assumption of constant variance. Treatments were however highly significantly different, and this is not changed by the transformation. (Remember Bartlett's test not very sensitive.)

Table I Student's *t*-distribution—values exceeded in two-tailed test with probability *P*

d.f.	$P=0\cdot1$	0·05	0·02	0·01	0·002	0·001
1	6·314	12·706	31·821	63·657	318·31	636·62
2	2·920	4·303	6·965	9·925	22·327	31·598
3	2·353	3·182	4·541	5·841	10·214	12·924
4	2·132	2·776	3·747	4·604	7·173	8·610
5	2·015	2·571	3·365	4·032	5·893	6·869
6	1·943	2·447	3·143	3·707	5·208	5·959
7	1·895	2·365	2·998	3·499	4·785	5·408
8	1·860	2·306	2·896	3·355	4·501	5·041
9	1·833	2·262	2·821	3·250	4·297	4·781
10	1·812	2·228	2·764	3·169	4·144	4·587
11	1·796	2·201	2·718	3·106	4·025	4·437
12	1·782	2·179	2·681	3·055	3·930	4·318
13	1·771	2·160	2·650	3·012	3·852	4·221
14	1·761	2·145	2·624	2·977	3·787	4·140
15	1·753	2·131	2·602	2·947	3·733	4·073
16	1·746	2·120	2·583	2·921	3·686	4·015
17	1·740	2·110	2·567	2·898	3·646	3·965
18	1·734	2·101	2·552	2·878	3·610	3·922
19	1·729	2·093	2·539	2·861	3·579	3·883
20	1·725	2·086	2·528	2·845	3·552	3·850
21	1·721	2·080	2·518	2·831	3·527	3·819
22	1·717	2·074	2·508	2·819	3·505	3·792
23	1·714	2·069	2·500	2·807	3·485	3·767
24	1·711	2·064	2·492	2·797	3·467	3·745
25	1·708	2·060	2·485	2·787	3·450	3·725
26	1·706	2·056	2·479	2·779	3·435	3·707
27	1·703	2·052	2·473	2·771	3·421	3·690
28	1·701	2·048	2·467	2·763	3·408	3·674
29	1·699	2·045	2·462	2·756	3·396	3·659
30	1·697	2·042	2·457	2·750	3·385	3·646
40	1·684	2·021	2·423	2·704	3·307	3·551
60	1·671	2·000	2·390	2·660	3·232	3·460
120	1·658	1·980	2·358	2·617	3·160	3·373
∞	1·645	1·960	2·326	2·576	3·090	3·291

The last row of the table (∞) gives values of *d*, the unit (standard) normal deviate.

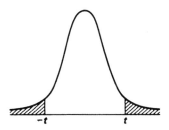

P is the sum of the two shaded areas

Table II   Values of the χ^2 distribution exceeded with probability P

P d.f.	0·995	0·975	0·050	0·025	0·010	0·005	0·001
1	392704.10⁻¹⁰	982069.10⁻⁹	3·84146	5·02389	6·63490	7·87944	10·828
2	0·0100251	0·0506356	5·99146	7·37776	9·21034	10·5966	13·816
3	0·0717218	0·215795	7·81473	9·34840	11·3449	12·8382	16·266
4	0·206989	0·484419	9·48773	11·1433	13·2767	14·8603	18·467
5	0·411742	0·831212	11·0705	12·8325	15·0863	16·7496	20·515
6	0·675727	1·23734	12·5916	14·4494	16·8119	18·5476	22·458
7	0·989256	1·68987	14·0671	16·0128	18·4753	20·2777	24·322
8	1·34441	2·17973	15·5073	17·5345	20·0902	21·9550	26·125
9	1·73493	2·70039	16·9190	19·0228	21·6660	23·5894	27·877
10	2·15586	3·24697	18·3070	20·4832	23·2093	25·1882	29·588
11	2·60322	3·81575	19·6751	21·9200	24·7250	26·7568	31·264
12	3·07382	4·40379	21·0261	23·3367	26·2170	28·2995	32·909
13	3·56503	5·00875	22·3620	24·7356	27·6882	29·8195	34·528
14	4·07467	5·62873	23·6848	26·1189	29·1412	31·3194	36·123
15	4·60092	6·26214	24·9958	27·4884	30·5779	32·8013	37·697
16	5·14221	6·90766	26·2962	28·8454	31·9999	34·2672	39·252
17	5·69722	7·56419	27·5871	30·1910	33·4087	35·7185	40·790
18	6·26480	8·23075	28·8693	31·5264	34·8053	37·1565	42·312
19	6·84397	8·90652	30·1435	32·8523	36·1909	38·5823	43·820
20	7·43384	9·59078	31·4104	34·1696	37·5662	39·9968	45·315
21	8·03365	10·28293	32·6706	35·4789	38·9322	41·4011	46·797
22	8·64272	10·9823	33·9244	36·7807	40·2894	42·7957	48·268
23	9·26043	11·6886	35·1725	38·0756	41·6384	44·1813	49·728
24	9·88623	12·4012	36·4150	39·3641	42·9798	45·5585	51·179
25	10·5197	13·1197	37·6525	40·6465	44·3141	46·9279	52·618
26	11·1602	13·8439	38·8851	41·9232	45·6417	48·2899	54·052
27	11·8076	14·5734	40·1133	43·1945	46·9629	49·6449	55·476
28	12·4613	15·3079	41·3371	44·4608	48·2782	50·9934	56·892
29	13·1211	16·0471	42·5570	45·7223	49·5879	52·3356	58·301
30	13·7867	16·7908	43·7730	46·9792	50·8922	53·6720	59·703
40	20·7065	24·4330	55·7585	59·3417	63·6907	66·7660	73·402
50	27·9907	32·3574	67·5048	71·4202	76·1539	79·4900	86·661
60	35·5345	40·4817	79·0819	83·2977	88·3794	91·9517	99·607
70	43·2752	48·7576	90·5312	95·0232	100·425	104·215	112·317
80	51·1719	57·1532	101·879	106·629	112·329	116·321	124·839
90	59·1963	65·6466	113·145	118·136	124·116	128·299	137·208
100	67·3276	74·2219	124·342	129·561	135·807	140·169	149·449

For d.f. > 100, test $\sqrt{2\chi^2_{(f)}}$ as $N\,(\sqrt{2f-1},\, 1)$.

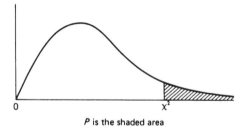

P is the shaded area

Table III Table of *F*-distribution

Upper 5% points

$\nu_2 \backslash \nu_1$	1	2	3	4	5	6	7	8	9	10	12	15	20	24	30	40	60	120	∞
1	161.4	199.5	215.7	224.6	230.2	234.0	236.8	238.9	240.5	241.9	243.9	245.9	248.0	249.1	250.1	251.1	252.2	253.3	254.3
2	18.51	19.00	19.16	19.25	19.30	19.33	19.35	19.37	19.38	19.40	19.41	19.43	19.45	19.45	19.46	19.47	19.48	19.49	19.50
3	10.13	9.55	9.28	9.12	9.01	8.94	8.89	8.85	8.81	8.79	8.74	8.70	8.66	8.64	8.62	8.59	8.57	8.55	8.53
4	7.71	6.94	6.59	6.39	6.26	6.16	6.09	6.04	6.00	5.96	5.91	5.86	5.80	5.77	5.75	5.72	5.69	5.66	5.63
5	6.61	5.79	5.41	5.19	5.05	4.95	4.88	4.82	4.77	4.74	4.68	4.62	4.56	4.53	4.50	4.46	4.43	4.40	4.36
6	5.99	5.14	4.76	4.53	4.39	4.28	4.21	4.15	4.10	4.06	4.00	3.94	3.87	3.84	3.81	3.77	3.74	3.70	3.67
7	5.59	4.74	4.35	4.12	3.97	3.87	3.79	3.73	3.68	3.64	3.57	3.51	3.44	3.41	3.38	3.34	3.30	3.27	3.23
8	5.32	4.46	4.07	3.84	3.69	3.58	3.50	3.44	3.39	3.35	3.28	3.22	3.15	3.12	3.08	3.04	3.01	2.97	2.93
9	5.12	4.26	3.86	3.63	3.48	3.37	3.29	3.23	3.18	3.14	3.07	3.01	2.94	2.90	2.86	2.83	2.79	2.75	2.71
10	4.96	4.10	3.71	3.48	3.33	3.22	3.14	3.07	3.02	2.98	2.91	2.85	2.77	2.74	2.70	2.66	2.62	2.58	2.54
11	4.84	3.98	3.59	3.36	3.20	3.09	3.01	2.95	2.90	2.85	2.79	2.72	2.65	2.61	2.57	2.53	2.49	2.45	2.40
12	4.75	3.89	3.49	3.26	3.11	3.00	2.91	2.85	2.80	2.75	2.69	2.62	2.54	2.51	2.47	2.43	2.38	2.34	2.30
13	4.67	3.81	3.41	3.18	3.03	2.92	2.83	2.77	2.71	2.67	2.60	2.53	2.46	2.42	2.38	2.34	2.30	2.25	2.21
14	4.60	3.74	3.34	3.11	2.96	2.85	2.76	2.70	2.65	2.60	2.53	2.46	2.39	2.35	2.31	2.27	2.22	2.18	2.13
15	4.54	3.68	3.29	3.06	2.90	2.79	2.71	2.64	2.59	2.54	2.48	2.40	2.33	2.29	2.25	2.20	2.16	2.11	2.07
16	4.49	3.63	3.24	3.01	2.85	2.74	2.66	2.59	2.54	2.49	2.42	2.35	2.28	2.24	2.19	2.15	2.11	2.06	2.01
17	4.45	3.59	3.20	2.96	2.81	2.70	2.61	2.55	2.49	2.45	2.38	2.31	2.23	2.19	2.15	2.10	2.06	2.01	1.96
18	4.41	3.55	3.16	2.93	2.77	2.66	2.58	2.51	2.46	2.41	2.34	2.27	2.19	2.15	2.11	2.06	2.02	1.97	1.92
19	4.38	3.52	3.13	2.90	2.74	2.63	2.54	2.48	2.42	2.38	2.31	2.23	2.16	2.11	2.07	2.03	1.98	1.93	1.88
20	4.35	3.49	3.10	2.87	2.71	2.60	2.51	2.45	2.39	2.35	2.28	2.20	2.12	2.08	2.04	1.99	1.95	1.90	1.84
21	4.32	3.47	3.07	2.84	2.68	2.57	2.49	2.42	2.37	2.32	2.25	2.18	2.10	2.05	2.01	1.96	1.92	1.87	1.81
22	4.30	3.44	3.05	2.82	2.66	2.55	2.46	2.40	2.34	2.30	2.23	2.15	2.07	2.03	1.98	1.94	1.89	1.84	1.78
23	4.28	3.42	3.03	2.80	2.64	2.53	2.44	2.37	2.32	2.27	2.20	2.13	2.05	2.01	1.96	1.91	1.86	1.81	1.76
24	4.26	3.40	3.01	2.78	2.62	2.51	2.42	2.36	2.30	2.25	2.18	2.11	2.03	1.98	1.94	1.89	1.84	1.79	1.73
25	4.24	3.39	2.99	2.76	2.60	2.49	2.40	2.34	2.28	2.24	2.16	2.09	2.01	1.96	1.92	1.87	1.82	1.77	1.71
26	4.23	3.37	2.98	2.74	2.59	2.47	2.39	2.32	2.27	2.22	2.15	2.07	1.99	1.95	1.90	1.85	1.80	1.75	1.69
27	4.21	3.35	2.96	2.73	2.57	2.46	2.37	2.31	2.25	2.20	2.13	2.06	1.97	1.93	1.88	1.84	1.79	1.73	1.67
28	4.20	3.34	2.95	2.71	2.56	2.45	2.36	2.29	2.24	2.19	2.12	2.04	1.96	1.91	1.87	1.82	1.77	1.71	1.65
29	4.18	3.33	2.93	2.70	2.55	2.43	2.35	2.28	2.22	2.18	2.10	2.03	1.94	1.90	1.85	1.81	1.75	1.70	1.64
30	4.17	3.32	2.92	2.69	2.53	2.42	2.33	2.27	2.21	2.16	2.09	2.01	1.93	1.89	1.84	1.79	1.74	1.68	1.62
40	4.08	3.23	2.84	2.61	2.45	2.34	2.25	2.18	2.12	2.08	2.00	1.92	1.84	1.79	1.74	1.69	1.64	1.58	1.51
60	4.00	3.15	2.76	2.53	2.37	2.25	2.17	2.10	2.04	1.99	1.92	1.84	1.75	1.70	1.65	1.59	1.53	1.47	1.39
120	3.92	3.07	2.68	2.45	2.29	2.17	2.09	2.02	1.96	1.91	1.83	1.75	1.66	1.61	1.55	1.50	1.43	1.35	1.25
∞	3.84	3.00	2.60	2.37	2.21	2.10	2.01	1.94	1.88	1.83	1.75	1.67	1.57	1.52	1.46	1.39	1.32	1.22	1.00

ν_1, ν_2 are upper, lower d.f. respectively.

Table III   (cont.)

Upper 2·5% points

$\nu_2 \backslash \nu_1$	1	2	3	4	5	6	7	8	9	10	12	15	20	24	30	40	60	120	∞
1	647·8	799·5	864·2	899·6	921·8	937·1	948·2	956·7	963·3	968·6	976·7	984·9	993·1	997·2	1001	1006	1010	1014	1018
2	38·51	39·00	39·17	39·25	39·30	39·33	39·36	39·37	39·39	39·40	39·41	39·43	39·45	39·46	39·46	39·47	39·48	39·49	39·50
3	17·44	16·04	15·44	15·10	14·88	14·73	14·62	14·54	14·47	14·42	14·34	14·25	14·17	14·12	14·08	14·04	13·99	13·95	13·90
4	12·22	10·65	9·98	9·60	9·36	9·20	9·07	8·98	8·90	8·84	8·75	8·66	8·56	8·51	8·46	8·41	8·36	8·31	8·26
5	10·01	8·43	7·76	7·39	7·15	6·98	6·85	6·76	6·68	6·62	6·52	6·43	6·33	6·28	6·23	6·18	6·12	6·07	6·02
6	8·81	7·26	6·60	6·23	5·99	5·82	5·70	5·60	5·52	5·46	5·37	5·27	5·17	5·12	5·07	5·01	4·96	4·90	4·85
7	8·07	6·54	5·89	5·52	5·29	5·12	4·99	4·90	4·82	4·76	4·67	4·57	4·47	4·42	4·36	4·31	4·25	4·20	4·14
8	7·57	6·06	5·42	5·05	4·82	4·65	4·53	4·43	4·36	4·30	4·20	4·10	4·00	3·95	3·89	3·84	3·78	3·73	3·67
9	7·21	5·71	5·08	4·72	4·48	4·32	4·20	4·10	4·03	3·96	3·87	3·77	3·67	3·61	3·56	3·51	3·45	3·39	3·33
10	6·94	5·46	4·83	4·47	4·24	4·07	3·95	3·85	3·78	3·72	3·62	3·52	3·42	3·37	3·31	3·26	3·20	3·14	3·08
11	6·72	5·26	4·63	4·28	4·04	3·88	3·76	3·66	3·59	3·53	3·43	3·33	3·23	3·17	3·12	3·06	3·00	2·94	2·88
12	6·55	5·10	4·47	4·12	3·89	3·73	3·61	3·51	3·44	3·37	3·28	3·18	3·07	3·02	2·96	2·91	2·85	2·79	2·72
13	6·41	4·97	4·35	4·00	3·77	3·60	3·48	3·39	3·31	3·25	3·15	3·05	2·95	2·89	2·84	2·78	2·72	2·66	2·60
14	6·30	4·86	4·24	3·89	3·66	3·50	3·38	3·29	3·21	3·15	3·05	2·95	2·84	2·79	2·73	2·67	2·61	2·55	2·49
15	6·20	4·77	4·15	3·80	3·58	3·41	3·29	3·20	3·12	3·06	2·96	2·86	2·76	2·70	2·64	2·59	2·52	2·46	2·40
16	6·12	4·69	4·08	3·73	3·50	3·34	3·22	3·12	3·05	2·99	2·89	2·79	2·68	2·63	2·57	2·51	2·45	2·38	2·32
17	6·04	4·62	4·01	3·66	3·44	3·28	3·16	3·06	2·98	2·92	2·82	2·72	2·62	2·56	2·50	2·44	2·38	2·32	2·25
18	5·98	4·56	3·95	3·61	3·38	3·22	3·10	3·01	2·93	2·87	2·77	2·67	2·56	2·50	2·44	2·38	2·32	2·26	2·19
19	5·92	4·51	3·90	3·56	3·33	3·17	3·05	2·96	2·88	2·82	2·72	2·62	2·51	2·45	2·39	2·33	2·27	2·20	2·13
20	5·87	4·46	3·86	3·51	3·29	3·13	3·01	2·91	2·84	2·77	2·68	2·57	2·46	2·41	2·35	2·29	2·22	2·16	2·09
21	5·83	4·42	3·82	3·48	3·25	3·09	2·97	2·87	2·80	2·73	2·64	2·53	2·42	2·37	2·31	2·25	2·18	2·11	2·04
22	5·79	4·38	3·78	3·44	3·22	3·05	2·93	2·84	2·76	2·70	2·60	2·50	2·39	2·33	2·27	2·21	2·14	2·08	2·00
23	5·75	4·35	3·75	3·41	3·18	3·02	2·90	2·81	2·73	2·67	2·57	2·47	2·36	2·30	2·24	2·18	2·11	2·04	1·97
24	5·72	4·32	3·72	3·38	3·15	2·99	2·87	2·78	2·70	2·64	2·54	2·44	2·33	2·27	2·21	2·15	2·08	2·01	1·94
25	5·69	4·29	3·69	3·35	3·13	2·97	2·85	2·75	2·68	2·61	2·51	2·41	2·30	2·24	2·18	2·12	2·05	1·98	1·91
26	5·66	4·27	3·67	3·33	3·10	2·94	2·82	2·73	2·65	2·59	2·49	2·39	2·28	2·22	2·16	2·09	2·03	1·95	1·88
27	5·63	4·24	3·65	3·31	3·08	2·92	2·80	2·71	2·63	2·57	2·47	2·36	2·25	2·19	2·13	2·07	2·00	1·93	1·85
28	5·61	4·22	3·63	3·29	3·06	2·90	2·78	2·69	2·61	2·55	2·45	2·34	2·23	2·17	2·11	2·05	1·98	1·91	1·83
29	5·59	4·20	3·61	3·27	3·04	2·88	2·76	2·67	2·59	2·53	2·43	2·32	2·21	2·15	2·09	2·03	1·96	1·89	1·81
30	5·57	4·18	3·59	3·25	3·03	2·87	2·75	2·65	2·57	2·51	2·41	2·31	2·20	2·14	2·07	2·01	1·94	1·87	1·79
40	5·42	4·05	3·46	3·13	2·90	2·74	2·62	2·53	2·45	2·39	2·29	2·18	2·07	2·01	1·94	1·88	1·80	1·72	1·64
60	5·29	3·93	3·34	3·01	2·79	2·63	2·51	2·41	2·33	2·27	2·17	2·06	1·94	1·88	1·82	1·74	1·67	1·58	1·48
120	5·15	3·80	3·23	2·89	2·67	2·52	2·39	2·30	2·22	2·16	2·05	1·94	1·82	1·76	1·69	1·61	1·53	1·43	1·31
∞	5·02	3·69	3·12	2·79	2·57	2·41	2·29	2·19	2·11	2·05	1·94	1·83	1·71	1·64	1·57	1·48	1·39	1·27	1·00

ν_1, ν_2 are upper, lower d.f. respectively.

Table III (cont.)

Upper 1% points

ν_2 \ ν_1	1	2	3	4	5	6	7	8	9	10	12	15	20	24	30	40	60	120	∞
1	4052	4999·5	5403	5625	5764	5859	5928	5981	6022	6056	6106	6157	6209	6235	6261	6287	6313	6339	6366
2	98·50	99·00	99·17	99·25	99·30	99·33	99·37	99·37	99·39	99·40	99·42	99·43	99·45	99·46	99·47	99·47	99·48	99·49	99·50
3	34·12	30·82	29·46	28·71	28·24	27·91	27·67	27·49	27·35	27·23	27·05	26·87	26·69	26·60	26·50	26·41	26·32	26·22	26·13
4	21·20	18·00	16·69	15·98	15·52	15·21	14·98	14·80	14·66	14·55	14·37	14·20	14·02	13·93	13·84	13·75	13·65	13·56	13·46
5	16·26	13·27	12·06	11·39	10·97	10·67	10·46	10·29	10·16	10·05	9·89	9·72	9·55	9·47	9·38	9·29	9·20	9·11	9·02
6	13·75	10·92	9·78	9·15	8·75	8·47	8·26	8·10	7·98	7·87	7·72	7·56	7·40	7·31	7·23	7·14	7·06	6·97	6·88
7	12·25	9·55	8·45	7·85	7·46	7·19	6·99	6·84	6·72	6·62	6·47	6·31	6·16	6·07	5·99	5·91	5·82	5·74	5·65
8	11·26	8·65	7·59	7·01	6·63	6·37	6·18	6·03	5·91	5·81	5·67	5·52	5·36	5·28	5·20	5·12	5·03	4·95	4·86
9	10·56	8·02	6·99	6·42	6·06	5·80	5·61	5·47	5·35	5·26	5·11	4·96	4·81	4·73	4·65	4·57	4·48	4·40	4·31
10	10·04	7·56	6·55	5·99	5·64	5·39	5·20	5·06	4·94	4·85	4·71	4·56	4·41	4·33	4·25	4·17	4·08	4·00	3·91
11	9·65	7·21	6·22	5·67	5·32	5·07	4·89	4·74	4·63	4·54	4·40	4·25	4·10	4·02	3·94	3·86	3·78	3·69	3·60
12	9·33	6·93	5·95	5·41	5·06	4·82	4·64	4·50	4·39	4·30	4·16	4·01	3·86	3·78	3·70	3·62	3·54	3·45	3·36
13	9·07	6·70	5·74	5·21	4·86	4·62	4·44	4·30	4·19	4·10	3·96	3·82	3·66	3·59	3·51	3·43	3·34	3·25	3·17
14	8·86	6·51	5·56	5·04	4·69	4·46	4·28	4·14	4·03	3·94	3·80	3·66	3·51	3·43	3·35	3·27	3·18	3·09	3·00
15	8·68	6·36	5·42	4·89	4·56	4·32	4·14	4·00	3·89	3·80	3·67	3·52	3·37	3·29	3·21	3·13	3·05	2·96	2·87
16	8·53	6·23	5·29	4·77	4·44	4·20	4·03	3·89	3·78	3·69	3·55	3·41	3·26	3·18	3·10	3·02	2·93	2·84	2·75
17	8·40	6·11	5·18	4·67	4·34	4·10	3·93	3·79	3·68	3·59	3·46	3·31	3·16	3·08	3·00	2·92	2·83	2·75	2·65
18	8·29	6·01	5·09	4·58	4·25	4·01	3·84	3·71	3·60	3·51	3·37	3·23	3·08	3·00	2·92	2·84	2·75	2·66	2·57
19	8·18	5·93	5·01	4·50	4·17	3·94	3·77	3·63	3·52	3·43	3·30	3·15	3·00	2·92	2·84	2·76	2·67	2·58	2·49
20	8·10	5·85	4·94	4·43	4·10	3·87	3·70	3·56	3·46	3·37	3·23	3·09	2·94	2·86	2·78	2·69	2·61	2·52	2·42
21	8·02	5·78	4·87	4·37	4·04	3·81	3·64	3·51	3·40	3·31	3·17	3·03	2·88	2·80	2·72	2·64	2·55	2·46	2·36
22	7·95	5·72	4·82	4·31	3·99	3·76	3·59	3·45	3·35	3·26	3·12	2·98	2·83	2·75	2·67	2·58	2·50	2·40	2·31
23	7·88	5·66	4·76	4·26	3·94	3·71	3·54	3·41	3·30	3·21	3·07	2·93	2·78	2·70	2·62	2·54	2·45	2·35	2·26
24	7·82	5·61	4·72	4·22	3·90	3·67	3·50	3·36	3·26	3·17	3·03	2·89	2·74	2·66	2·58	2·49	2·40	2·31	2·21
25	7·77	5·57	4·68	4·18	3·85	3·63	3·46	3·32	3·22	3·13	2·99	2·85	2·70	2·62	2·54	2·45	2·36	2·27	2·17
26	7·72	5·53	4·64	4·14	3·82	3·59	3·42	3·29	3·18	3·09	2·96	2·81	2·66	2·58	2·50	2·42	2·33	2·23	2·13
27	7·68	5·49	4·60	4·11	3·78	3·56	3·39	3·26	3·15	3·06	2·93	2·78	2·63	2·55	2·47	2·38	2·29	2·20	2·10
28	7·64	5·45	4·57	4·07	3·75	3·53	3·36	3·23	3·12	3·03	2·90	2·75	2·60	2·52	2·44	2·35	2·26	2·17	2·06
29	7·60	5·42	4·54	4·04	3·73	3·50	3·33	3·20	3·09	3·00	2·87	2·73	2·57	2·49	2·41	2·33	2·23	2·14	2·03
30	7·56	5·39	4·51	4·02	3·70	3·47	3·30	3·17	3·07	2·98	2·84	2·70	2·55	2·47	2·39	2·30	2·21	2·11	2·01
40	7·31	5·18	4·31	3·83	3·51	3·29	3·12	2·99	2·89	2·80	2·66	2·52	2·37	2·29	2·20	2·11	2·02	1·92	1·80
60	7·08	4·98	4·13	3·65	3·34	3·12	2·95	2·82	2·72	2·63	2·50	2·35	2·20	2·12	2·03	1·94	1·84	1·73	1·60
120	6·85	4·79	3·95	3·48	3·17	2·96	2·79	2·66	2·56	2·47	2·34	2·19	2·03	1·95	1·86	1·76	1·66	1·53	1·38
∞	6·63	4·61	3·78	3·32	3·02	2·80	2·64	2·51	2·41	2·32	2·18	2·04	1·88	1·79	1·70	1·59	1·47	1·32	1·00

ν_1, ν_2 are upper, lower d.f. respectively.

Table III　(cont.)

Upper 0·1% points

$v_2 \backslash v_1$	∞	120	60	40	30	24	20	15	12	10	9	8	7	6	5	4	3	2	1
1	6366*	6340*	6313*	6287*	6261*	6235*	6209*	6158*	6107*	6056*	6023*	5981*	5929*	5859*	5764*	5625*	5404*	5000*	4053*
2	999·5	999·5	999·5	999·5	999·5	999·5	999·4	999·4	999·4	999·4	999·4	999·4	999·4	999·3	999·3	999·2	999·2	999·0	998·5
3	123·5	124·0	124·5	125·0	125·4	125·9	126·4	127·4	128·3	129·2	129·9	130·6	131·6	132·8	134·6	137·1	141·1	148·5	167·0
4	44·08	44·40	44·75	45·09	45·43	45·77	46·10	46·76	47·41	48·05	48·47	49·00	49·66	50·53	51·71	53·44	56·18	61·25	74·14
5	23·79	24·06	24·33	24·60	24·87	25·14	25·39	25·91	26·42	26·92	27·24	27·64	28·16	28·84	29·75	31·09	33·20	37·12	47·18
6	15·75	15·99	16·21	16·44	16·67	16·89	17·12	17·56	17·99	18·41	18·69	19·03	19·46	20·03	20·81	21·92	23·70	27·00	35·51
7	11·70	11·91	12·12	12·33	12·53	12·73	12·93	13·32	13·71	14·08	14·33	14·63	15·02	15·52	16·21	17·19	18·77	21·69	29·25
8	9·33	9·53	9·73	9·92	10·11	10·30	10·48	10·84	11·19	11·54	11·77	12·04	12·40	12·86	13·49	14·39	15·83	18·49	25·42
9	7·81	8·00	8·19	8·37	8·55	8·72	8·90	9·24	9·57	9·89	10·11	10·37	10·70	11·13	11·71	12·56	13·90	16·39	22·86
10	6·76	6·94	7·12	7·30	7·47	7·64	7·80	8·13	8·45	8·75	8·96	9·20	9·52	9·92	10·48	11·28	12·55	14·91	21·04
11	6·00	6·17	6·35	6·52	6·68	6·85	7·01	7·32	7·63	7·92	8·12	8·35	8·66	9·05	9·58	10·35	11·56	13·81	19·69
12	5·42	5·59	5·76	5·93	6·09	6·25	6·40	6·71	7·00	7·29	7·48	7·71	8·00	8·38	8·89	9·63	10·80	12·97	18·64
13	4·97	5·14	5·30	5·47	5·63	5·78	5·93	6·23	6·52	6·80	6·98	7·21	7·49	7·86	8·35	9·07	10·21	12·31	17·81
14	4·60	4·77	4·94	5·10	5·25	5·41	5·56	5·85	6·13	6·40	6·58	6·80	7·08	7·43	7·92	8·62	9·73	11·78	17·14
15	4·31	4·47	4·64	4·80	4·95	5·10	5·25	5·54	5·81	6·08	6·26	6·47	6·74	7·09	7·57	8·25	9·34	11·34	16·59
16	4·06	4·23	4·39	4·54	4·70	4·85	4·99	5·27	5·55	5·81	5·98	6·19	6·46	6·81	7·27	7·94	9·00	10·97	16·12
17	3·85	4·02	4·18	4·33	4·48	4·63	4·78	5·05	5·32	5·58	5·75	5·96	6·22	6·56	7·02	7·68	8·73	10·66	15·72
18	3·67	3·84	4·00	4·15	4·30	4·45	4·59	4·87	5·13	5·39	5·56	5·76	6·02	6·35	6·81	7·46	8·49	10·39	15·38
19	3·51	3·68	3·84	3·99	4·14	4·29	4·43	4·70	4·97	5·22	5·39	5·59	5·85	6·18	6·62	7·26	8·28	10·16	15·08
20	3·38	3·54	3·70	3·86	4·00	4·15	4·29	4·56	4·82	5·08	5·24	5·44	5·69	6·02	6·46	7·10	8·10	9·95	14·82
21	3·26	3·42	3·58	3·74	3·88	4·03	4·17	4·44	4·70	4·95	5·11	5·31	5·56	5·88	6·32	6·95	7·94	9·77	14·59
22	3·15	3·32	3·48	3·63	3·78	3·92	4·06	4·33	4·58	4·83	4·99	5·19	5·44	5·76	6·19	6·81	7·80	9·61	14·38
23	3·05	3·22	3·38	3·53	3·68	3·82	3·96	4·23	4·48	4·73	4·89	5·09	5·33	5·65	6·08	6·69	7·67	9·47	14·19
24	2·97	3·14	3·29	3·45	3·59	3·74	3·87	4·14	4·39	4·64	4·80	4·99	5·23	5·55	5·98	6·59	7·55	9·34	14·03
25	2·89	3·06	3·22	3·37	3·52	3·66	3·79	4·06	4·31	4·56	4·71	4·91	5·15	5·46	5·88	6·49	7·45	9·22	13·88
26	2·82	2·99	3·15	3·30	3·44	3·59	3·72	3·99	4·24	4·48	4·64	4·83	5·07	5·38	5·80	6·41	7·36	9·12	13·74
27	2·75	2·92	3·08	3·23	3·38	3·52	3·66	3·92	4·17	4·41	4·57	4·76	5·00	5·31	5·73	6·33	7·27	9·02	13·61
28	2·69	2·86	3·02	3·18	3·32	3·46	3·60	3·86	4·11	4·35	4·50	4·69	4·93	5·24	5·66	6·25	7·19	8·93	13·50
29	2·64	2·81	2·97	3·12	3·27	3·41	3·54	3·80	4·05	4·29	4·45	4·64	4·87	5·18	5·59	6·19	7·12	8·85	13·39
30	2·59	2·76	2·92	3·07	3·22	3·36	3·49	3·75	4·00	4·24	4·39	4·58	4·82	5·12	5·53	6·12	7·05	8·77	13·29
40	2·23	2·41	2·57	2·73	2·87	3·01	3·15	3·40	3·64	3·87	4·02	4·21	4·44	4·73	5·13	5·70	6·60	8·25	12·61
60	1·89	2·08	2·25	2·41	2·55	2·69	2·83	3·08	3·31	3·54	3·69	3·87	4·09	4·37	4·76	5·31	6·17	7·76	11·97
120	1·54	1·76	1·95	2·11	2·26	2·40	2·53	2·78	3·02	3·24	3·38	3·55	3·77	4·04	4·42	4·95	5·79	7·32	11·38
∞	1·00	1·45	1·66	1·84	1·99	2·13	2·27	2·51	2·74	2·96	3·10	3·27	3·47	3·74	4·10	4·62	5·42	6·91	10·83

v_1, v_2 are upper, lower d.f. respectively.　　* Multiply these entries by 100.

Table IV Values of the correlation coefficient, *r*, which differ significantly from 0 at the 5%, 1%, 0.1% levels, in a two-tail test

d.f.	0·05	0·01	0·001	d.f.	0·05	0·01	0·001
1	0·9^{2}692	0·9^{2}877	0·9^{2}877	16	0·468	0·590	0·708
2	·9500	·9^{2}000	·9^{2}000	17	·456	·575	·693
3	·878	·9587	·9^{2}114	18	·444	·561	·679
4	·811	·9172	·9741	19	·433	·549	·665
5	·754	·875	·9509	20	·423	·537	·652
6	0·707	0·834	0·9249	25	0·381	0·487	0·597
7	·666	·798	·898	30	·349	·449	·554
8	·632	·765	·872	35	·325	·418	·519
9	·602	·735	·847	40	·304	·393	·490
10	·576	·708	·823	45	·288	·372	·465
11	0·553	0·684	0·801	50	0·273	0·354	0·443
12	·532	·661	·780	60	·250	·325	·408
13	·514	·641	·760	70	·232	·302	·380
14	·497	·623	·742	80	·217	·283	·357
15	·482	·606	·725	90	·205	·267	·338
				100	·195	·254	·321

Note Degrees of freedom, d.f., are $(n-2)$ if *n* pairs of data (x, y) are available.

The foregoing Tables are reprinted, by permission of the Trustees, from *Biometrika Tables for Statisticians*, 3rd Edition (1966), ed. E. S. Pearson and H. O. Hartley.

Table V Values of Spearman's rank correlation coefficient r_s which differ significantly from 0 at the 5% and 1% levels using a two-tail test

Sample size *n*	5% level	1% level
5	1·000	–
6	0·886	1·000
7	0·786	0·929
8	0·738	0·881
9	0·683	0·833
10	0·648	0·794

For sample sizes greater than $n = 10$ use Table IV.

Table VI   Table of random digits

10582	30143	89214	52134	76280	77823	61674	96898	90487	43998
51753	56087	71524	64913	81706	33984	09919	86969	75553	87375
96050	08123	28557	04240	33606	10776	64239	81900	74880	92654
93998	95705	73353	26933	66089	25177	62387	34932	62021	34444
70974	45757	31830	09589	31037	91886	51780	21912	16444	52881
25833	71286	76375	43640	92551	46510	68950	60168	26399	04599
55060	28982	92650	71622	36740	05869	17828	29377	10120	90851
29436	79967	34383	85646	04715	80695	39283	50543	26875	94047
80180	08706	17875	72123	69723	52846	71310	72507	25702	33449
40842	32742	44671	72953	54811	39495	05023	61569	60805	26580
31481	16208	60372	94367	88977	35393	08681	53325	92547	31622
06045	35097	38319	17264	40640	63022	01496	28439	04197	63858
41446	12336	54072	47189	56085	25215	89943	41153	18496	76869
22301	07404	60943	75921	02932	50090	51949	86415	51919	98125
38199	09042	26771	15881	80204	61281	61610	24501	01935	33256
06273	93282	55034	79777	75241	11762	11274	41685	24117	98311
92201	02587	31599	27987	25678	69736	94487	41653	79550	92949
70782	80894	95413	36338	04237	19954	71137	23584	87069	10407
05245	40934	96832	33415	62058	87179	31542	18174	54711	21882
85607	45719	65640	33241	04852	87636	43840	42242	22092	28975

Table VII Values of z, the standard normal variable, from 0·0 by steps of 0·01 to 3·9, showing the cumulative probability up to z.
(Probability correct to four decimal places).

z	0·00	0·01	0·02	0·03	0·04	0·05	0·06	0·07	0·08	0·09
0·0	·5000	·5040	·5080	·5120	·5160	·5199	·5239	·5279	·5319	·5359
·1	·5398	·5438	·5478	·5517	·5557	·5596	·5636	·5675	·5714	·5753
·2	·5793	·5832	·5871	·5910	·5948	·5987	·6026	·6064	·6103	·6141
·3	·6179	·6217	·6255	·6293	·6331	·6368	·6406	·6443	·6480	·6517
·4	·6554	·6591	·6628	·6664	·6700	·6736	·6772	·6808	·6844	·6879
·5	·6915	·6950	·6985	·7019	·7054	·7088	·7123	·7157	·7190	·7224
·6	·7257	·7291	·7324	·7357	·7389	·7422	·7454	·7486	·7517	·7549
·7	·7580	·7611	·7642	·7673	·7704	·7734	·7764	·7794	·7823	·7852
·8	·7881	·7910	·7939	·7967	·7995	·8023	·8051	·8078	·8106	·8133
·9	·8159	·8186	·8212	·8238	·8264	·8289	·8315	·8340	·8365	·8389
1·0	·8413	·8438	·8461	·8485	·8508	·8531	·8554	·8577	·8599	·8621
·1	·8643	·8665	·8686	·8708	·8729	·8749	·8770	·8790	·8810	·8830
·2	·8849	·8869	·8888	·8907	·8925	·8944	·8962	·8980	·8997	·9015
·3	·9032	·9049	·9066	·9082	·9099	·9115	·9131	·9147	·9162	·9177
·4	·9192	·9207	·9222	·9236	·9251	·9265	·9279	·9292	·9306	·9319
·5	·9332	·9345	·9357	·9370	·9382	·9394	·9406	·9418	·9429	·9441
·6	·9452	·9463	·9474	·9484	·9495	·9505	·9515	·9525	·9535	·9545
·7	·9554	·9564	·9573	·9582	·9591	·9599	·9608	·9616	·9625	·9633
·8	·9641	·9649	·9656	·9664	·9671	·9678	·9686	·9693	·9699	·9706
·9	·9713	·9719	·9726	·9732	·9738	·9744	·9750	·9756	·9761	·9767
2·0	·9772	·9778	·9783	·9788	·9793	·9798	·9803	·9808	·9812	·9817
·1	·9821	·9826	·9830	·9834	·9838	·9842	·9846	·9850	·9854	·9857
·2	·9861	·9864	·9868	·9871	·9875	·9878	·9881	·9884	·9887	·9890
·3	·9893	·9896	·9898	·9901	·9904	·9906	·9909	·9911	·9913	·9916
·4	·9918	·9920	·9922	·9925	·9927	·9929	·9931	·9932	·9934	·9936
·5	·9938	·9940	·9941	·9943	·9945	·9946	·9948	·9949	·9951	·9952
·6	·9953	·9955	·9956	·9957	·9959	·9960	·9961	·9962	·9963	·9964
·7	·9965	·9966	·9967	·9968	·9969	·9970	·9971	·9972	·9973	·9974
·8	·9974	·9975	·9976	·9977	·9977	·9978	·9979	·9979	·9980	·9981
·9	·9981	·9982	·9982	·9983	·9984	·9984	·9985	·9985	·9986	·9986
3·0	·9987	·9987	·9987	·9988	·9988	·9989	·9989	·9989	·9990	·9990
·1	·9990	·9991	·9991	·9991	·9992	·9992	·9992	·9992	·9993	·9993
·2	·9993	·9993	·9994	·9994	·9994	·9994	·9994	·9995	·9995	·9995
·3	·9995	·9995	·9995	·9996	·9996	·9996	·9996	·9996	·9996	·9997
·4	·9997	·9997	·9997	·9997	·9997	·9997	·9997	·9997	·9997	·9998
·5	·9998	·9998	·9998	·9998	·9998	·9998	·9998	·9998	·9998	·9998
·6	·9998	·9998	·9999	·9999	·9999	·9999	·9999	·9999	·9999	·9999
·7	·9999	·9999	·9999	·9999	·9999	·9999	·9999	·9999	·9999	·9999
·8	·9999	·9999	·9999	·9999	·9999	·9999	·9999	·9999	·9999	·9999
·9	1·0000									

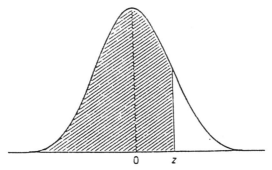

The curve is $N(0, 1)$, the standard normal variable. The table entry is the shaded area $\Phi(z) = \Pr(Z < z)$. For example, when $z = 1·96$ the shaded area is 0·9750. Critical values of the standard normal distribution will be found in the bottom row of Table I.

Index